The Age of Systems

THE AGE OF SYSTEMS

The Human Dilemma

William Exton, Jr.

American Management Association

133675

658.31
E 96

© American Management Association, Inc., 1972.
All rights reserved. Printed in the United States of America.

This publication may not be reproduced, stored in a retrieval system, or transmitted in whole or in part, in any form or by any means, electronic, mechanical, photocopying, recording, or otherwise, without the prior written permission of the Association.

International standard book number: 0-8144-5282-5
Library of Congress catalog card number: 72-173316

First printing

TO Florence Nightingale August Phillips Exton
with a fortunate son's love;
and with everlasting gratitude
for the very large proportion
of her more than ninety years
of which he has been a prime beneficiary.

Contents

Foreword

THE age in which we are living has been called the Age of Communication, and of Psychology; of Electronics and of Pollution; of Anticolonialism and of Multinational Corporations; of Technology and of the Emerging Underdog; of Space Exploration and of the Generation Gap; of Anxiety; and even of Aquarius. But in the inevitable perspective of time it will probably be known as the Age of Systems.

There is nothing so distinctive, so special and unprecedented, so significant and so basically emergent in our age as the growth, development, and proliferation of systematization; the prevalence of systems for all purposes; the recourse to systematization as the fundamental characteristic of projects and the solution to problems. Systems represent and contribute to progress, economy, efficiency, improvement, benefits, advantages, and many other generally good things. Systems are based on the unalterable logic of finding a good way to accomplish a desired purpose (usually on a continuous or repetitive basis) and then doing it faithfully and consistently in that way.

But this age is not only the Age of Systems. It is also an age in which many other developments are taking place, and some of them are extremely unwelcome. These include the behavior of impressively large groups in our most advanced economies. Among the most disturbing developments are the apparent rejection of their elders by youth; the violent, destructive activism of some dissidents, students, blacks, and a few members of other minority groups; the uncontrolled spread of drug abuse; and the frightening rise in lawbreaking and disorder—the appalling growth of major and minor crime.

Since the steepest upcurve in the onset of these most undesirable phenomena coincides with the exponential proliferation of systems, from which we hope for so much, we should ask ourselves if there is some connection or relationship between these parallel developments; and if so, what elements of either one may be causative of or contributory to the other; and what we can or should do about it.

The probability of some connection is strengthened by the fact that these concurrent trends seem to involve and overlap with many of the same individuals, since the staffing of many systems is drawn largely from the young—and these are students, recent graduates, or those involved with students; often those entering upon their first regular employment; and also increasingly from among the rising generation of blacks and other ethnic groups.

Since systems are created by technicians and professional specialists acting under the direction of entrepreneurs and senior management, and since the contact of most employees (and others) with a system can occur only after its formation, it would appear that a primary causative relation would not be likely to flow from the human involvement toward the system. (This is not to deny the obvious reactive effects.) On the other hand, contact with the system as an employee and participant is a most significant event in the lives of many adolescents and young adults; increasingly, the opportunities for employment available to high school graduates are found among the lowest echelons of large system operations. It may be assumed, then, that a primary causative relation may likely derive from the system, acting upon its individual recruits.

Those who operate systems, then, may be considered to have some degree of moral responsibility for system effects upon their young employees. But real and effective concern for the effects of system involvement upon employees, regrettably, is not often observed among those who design or operate systems.

In fact, the opposite is quite general: The concern of many managers is for what they consider to be the effects of the employees upon the system, rather than the reverse. Most managements are seriously troubled about the inadequate or inappropriate behavior of many of their employees, since managements face grave difficulties in operating their systems as planned when employees fail to perform according to plan.

And yet the patterns of behavior of employees—especially young employees—within systems are eloquent of the causes that underly and give rise to such behavior.

No genius is required to understand the underlying cause of absenteeism. The absent ones would obviously rather be elsewhere—perhaps anywhere—than at work. This is clear evidence of a very strong aversion to the work place.

No special insight is needed to appreciate the cause of low-quality output. The worker places no value on the results of his work because he is totally lacking in personal involvement in his job.

No deep wisdom is necessary to know the cause of the failure of discipline and the all-too-common lack of respect for foremen, supervisors, and managers. There is clear evidence of a gulf between those who value the system and those who do not.

In the light of such irrefutable evidence that many employees, in effect, appear to reject major features of their employment, we must ask why they stay on the job at all (granted the very high rate of attrition). And we come up with two answers: They see no alternative or preferable employment (they will quit like a shot if they do); and they need or want whatever they can earn.

So they consent to lend themselves for hire, like prostitutes, for a perfunctory performance, without interest in employer or results.

This condition inevitably teaches festering self-contempt; and also subjects its victims to excruciating boredom. Boredom renders these employees vulnerable to alternatives and distractions—often to violent reactions. Self-contempt abolishes standards ("I don't care!") and nullifies scruples and morals ("I've got nothing to lose!").

The more self-respect the individual has to begin with, the greater the resentment that must inevitably be engendered by finding himself in a situation in which he is valued no more highly than as an easily replaceable cog in a simple, purposeless, endless mechanism. If he aspires to promotion, he soon learns that he will have little opportunity to demonstrate his potential; and having no respect for his supervisor, he may have little desire to replace him in the same position. Naturally, he will look elsewhere; turnover is great. With no more attractive alternative, he stays—and grows bitter.

The consequences of such degenerative processes are not easily measured; but the discrepancies between the potential productivity of a system and its actual useful output—quality considered—would suggest at least one parameter from the point of view of management.

The psychic cost to the employees is not subject to measurement, but is at least slightly suggested by the great waste of the potential of intelligent, dynamic youngsters; and perhaps also by the extremes to which some adolescents and young adults feel themselves driven as soon as they are away from the work place.

If the attitudes of many young employees are antisystem, to what extent do they bring these attitudes with them, and to what extent does their participation in a system generate or reinforce these attitudes?

Much further and more profound study is needed before adequate answers can be given to such questions; but common observation provides something to go on. These observations relate to at least three levels: economic factors, cultural factors, and the character of the work. The first two of these have been widely discussed elsewhere; the third will be discussed here.

To consider the fundamental character of a system is to recognize the seemingly insoluble paradox between the vast, complex total operation simplified into an integrated set of processes requiring little but routine performance from their participants; and the too simple individual activities, infinitely repeated, thus required of some extremely complex human beings. The job and the person are, in fact, incompatible.

Those who design the system plan it to produce desired results in the "best" way. How the system is to operate determines, of course, the character of the component processes, and so also determines, in effect, the way in which each individual participant must operate. For each, there is no real alternative, even when constructively motivated.

No new thought is required as to the "how"—the system takes care of that. It is only necessary to fulfill the requirements of the system. These requirements are divided among relatively simple, standardized patterns of activity that are commonly designed to be readily conducted by groups or individuals with far less knowledge, ability, and skill than would be required without the system.

All systems rest on the establishment of a fixed, determined process or series of processes. Most processes also rest upon the provision of equipment—usually adapted and highly specialized—without which the processes would not be feasible; or would, at least, be less efficiently accomplished.

A system counts upon the organization of the individual participants to fulfill a plan. A system may be as simple and improvised as the bucket brigade that enables the neighbors to help fight a fire better than by relying upon individual efforts to fetch and carry water to throw on the flames. And it may be as complex as the whole panoply of buildings and special equipment, alarms and communications, trained personnel and specified procedures that enable the modern fire department of a major city to carry out its parallel functions.

It is of the essence of a system that *what to do* is predetermined; so that individual initiative is generally superfluous in the operation of a system. It is also characteristic of systems that functions are assigned, responsibilities are delegated, roles are designated, activities are prescribed, so that the individual's permissible behavior is often quite rigidly limited in scope. The contribution of the individual is both prescribed and circumscribed.

The advantage to the individual presumably derives from the greater productivity of a system—especially when the system makes possible the use of highly efficient equipment. Under appropriate circumstances the individual can and will receive more for participating in a productive system than he could or would otherwise. (However, this is not always true; and the existence of a system may stand in the way of an individual acting in a manner that would be more advantageous to himself, as when a system replaces more exacting individualized activities.)

The inevitable result of all this is to place system and individuality in total opposition, so that the more perfected and advanced the system, the less allowed to or expected of the individual, and the more intelligent, dynamic, creative, and ambitious the individual, the more he will find himself frustrated by participation in the system.

We see, then, the apparently limitless progress of system design, development, and scope moving toward eliminating the kinds of experience that enable individuals to develop, and to find satisfaction in their work. And at the same time we see the social, economic, and cultural developments which lead people to demand more and greater satisfactions from their activities, and less regimentation and restraint from others—and especially from those whose values they reject.

This is a collision course.

To reject a system may not necessarily be to reject *the* system; but if a dynamic young person suffers a thoroughly unacceptable experience within the system that is a part of a well-known organization (often a large, nationally operating corporation) it can be readily understood if he assumes this to be typical of the experience that must be endured in *all* systems in *all* organizations. What may appear to be an unjustifiable overgeneralization is all too likely to be supported by the parallel evaluations of siblings and intimates, classmates and acquaintances.

He is applying *his* value system; and by his lights, his evaluation is inescapable. And so the rejection of a system leads to the rejection of "the system," and so, perhaps inevitably, to rejection of "the establishment."

And if, embittered, disillusioned, resentful, and hostile, he nonetheless remains within the system and strives for advancement, can we for a moment assume that his experience has been beneficial? Surely not, except in the very limited sense that one may learn from *any* experience; and that if one is to administer a system, one should first serve within it. While undeniably true, such truisms cannot justify all that must now be endured while making one's low-level contribution to a typical system.

What can be done?

It would appear that the development of systems and the course of a rising generation are hopelessly at loggerheads. But there are so many and such diverse factors and elements on both sides that surely some compromise, some adjustment, some degree of accommodation may be found; if not on a broad front, then at least here and there. Certainly more of an effort—and a more concerted and sophisticated and determined effort—should and must be made.

The Age of Systems is here to stay. So we must learn to live with it successfully; benefiting, but not paying too high a price. It is hoped that this book will be helpful in this essential endeavor.

Introduction

THE essential functions of most managers must increasingly be focused upon the development, operation, or exploitation of systems. As present system functions grow and are extended, additional organizational functions are systematized. Many systems become more and more elaborate, more completely mechanized, more automated. There is thus an accelerating increase, both quantitative and qualitative, in systematization of organizational activities.

Many advantages have been sought through systematization, and some have been realized to a gratifying degree. However, there have been many disappointments. Many expectations for system effectiveness have not been met. Some systems have been failures, and a large proportion of others seem to generate problems, difficulties, and costs that were not foreseen and are not easily eliminated.

Some failures of systems, of course, have been due to changed conditions, which may or may not have been predictable. Others have been caused by misapplication or inadequate functioning of equipment or other physical causes. But in all probability, a large majority of deficiencies in the operations of systems have been the result of deficiencies in human performance. There has been too great a disparity between planned human participation in the systems concept and actual human performance on the job.

Examples of such deficiencies, attributable to such disparities, have been distressingly characteristic of systems functioning in clerical operations—probably far more so, for instance, than of the more mechanically based systems characteristic of manufacturing processes.

Every system, as installed, is expected to fulfill a specified purpose, to perform a predetermined function, and to make a defined contribution to the overall functioning of the organization—more or less, and increasingly, integrated with other functions to constitute major systems. Under sound management doctrine, system operations are planned for specified minimum levels of productivity, within acceptable parameters of cost and of drain upon managerial and other resources of the organization.

When physical systems fail to meet such production standards, the problems are generally seen as falling within the purview of engineers of various kinds, quality control specialists, process experts, and others offering necessary capabilities in physical and mathematical disciplines. When clerical systems fall short of planned efficiency (and the problems call loudly enough for solution), management usually turns to methods and procedures specialists, setters of work standards, forms designers, and, where relevant, to equipment specialists. In either situation, human factors may be significant, even critical, and this fact may sometimes be adequately recognized. But consideration of human factors in specific situations has not, apparently, led to a sufficiently general, basic recognition of the quintessential nature of systems in relation to and their effects upon the human participants.

The advent of the computer has greatly emphasized these trends. The growth, in numbers and extent, of computerized information systems of all kinds has created a vastly greater prevalence of situations in which systems are designed around data-processing-equipment capabilities and the necessary or appropriate related processes and programs. Such situations involve the detailed, highly specialized clerical operations required to provide the inputs and to handle the outputs. And the roles assigned to people in such operations generally are, in fact, too often largely incompatible with the intangible value systems of the employees.

The employees face the unremitting requirements of daylong, assiduous concentration on what are, essentially, meaningless (to them) symbols, and of sustained care and attention to simple, basically repetitive tasks. For this, they are penalized by the absence both of inherent interest in the work and of satisfaction for achievement. These factors and others militate powerfully against, and overwhelmingly preclude, the psychological and social factors necessary for the support of excellent performance. Individual efficiency, in such situations, requires a highly adapted adjustment, which is seldom fully attained. As a result, the human contributions to such systems often constitute a serious, and sometimes intolerable, limit to system efficiency.

At the same time, employment in such system operations becomes so unattractive (to those unable or unwilling to adjust) that seekers of employment will prefer available alternatives, and such systems, therefore, tend to be staffed with those who, qualified or not, have no preferable alternatives. This selective effect is undoubtedly reflected in the relative paucity of managerial or even supervisory potential in many employee groups.

If systematization may be regarded as a medicine for curing organizations of inefficiencies, weaknesses, and ailments, then it should also be regarded as often having potent, undesirable side effects, the symptoms of which afflict the employees who man the systems. These side effects are increasingly prevalent among those who staff our business and governmental organizations. And the effects are often more important, more serious, and have longer-term, longer-range results than are generally realized.

To employ a somewhat different simile, systematization may be regarded as a strategy, but the introduction of each system involves tactics. These have specific, tactical objectives, which may or may not be fully realized. But in executing such tactics, those employed may suffer various degrees of psychic casualty. And the overall effect upon the organization, when a substantial proportion of the staff has been thus affected, may be quite deleteriously strategic.

In the course of his professional activities as a management consultant, the author has been impressed with both the tactical and the strategic considerations; that is, the way systems often fail to reach their full potential through the inadequacies of human contribution, and the deplorably adverse effects of participation in many systems operations upon the human component.

This book is the result of extensive consideration of the many interrelated, interlocked problems bearing upon human participation in and contribution to the functioning of systems, and of the implications of this for the long-term health and vigor of organizations.

It is hoped this book will be helpful in improving both the effectiveness of systems and the quality of life of those who must work within them. Only along these lines can the organizations of a free-enterprise system continue to maintain and to enlarge the efficiency, initiative, innovation, self-regenerating capabilities, psychic energy, and drive that have created the American economy.

1 Perspective on Systems

WHEN Commander Neil Armstrong took his "giant step for mankind" onto the surface of the moon, he was in himself symbolizing and manifesting the most remarkable example ever of man's use of systems. And when Susie Jones sits all day in an office of the telephone company, sorting paper charge slips for long-distance calls, she is exemplifying something like the opposite extreme—the way that systems use people.

The term *system* represents the principle of functional combination of resources to produce intended results or effects. The combination may be great or small, simple or enormously complex, active or potential, solitary or parallel, new or old, static or dynamic. The intended effect may be fixed or otherwise, unique or repetitive or continuous, geographically or spacially defined or unlimited in territorial scope, physical or symbolic, tangible or intangible.

A system may be made up of other interrelated systems, as is the body. A system may or may not involve people internally, as school systems involve teachers and pupils and telephone systems involve employees; and it may or may not involve people externally, as transportation systems carry freight and as communication systems transmit messages to and from people.

Characteristics of Systems

The terms *system* and *systematic* have been with us a long time. From the solar system to the Bell System, the word *system* connotes a regularity in functional relationships among the members of a group—whether heavenly bodies or telephones. Thus it tends to suggest related tangibles such as hardware, equipment, technology. The term is also applied to a sequence of related activities to be applied with a generally predictable or desired result, such as the Goren system for playing contract bridge or a system of taxation.

Common usage today extends the application of the term *system* from telephones and railways to other interrelated functioning complexes that combine facilities and required modes of operation for the production or attainment of desired results. These may be extremely complex and esoteric, such as the Polaris or Minuteman systems and other monumental military elaborations, or they may be relatively simple in principle but formidable in extent of elaboration, such as the collection and processing of the raw data of the census, or the operations of the New York Clearing House or a major school system.

Or they may be as familiar and unimpressive as the clerical procedures in many offices, the processing of materials in many industrial installations, the handling of goods in many warehouses; or they may be any other activity in which individuals are involved, but in which the nature and extent of their contributions are dictated by and subordinate to the requirements of the overall system. Most definitely included in this classification are many clerical operations, data processing activities, management information and control systems, and other situations where some individuals play a part in an ongoing organization activity without making any individual contribution to the results.

Even closer to home, we possess within ourselves a nervous system, a circulatory system, a lymphatic system, and other systems, all functioning together to support the life system. No one of these systems, however perfectly capable of functioning, can do so for long without adequate support from the others.

These illustrations suggest that the concepts to be conveyed by the term *system* involve a unified mixture of interconnected functions, parallel and/or successive, all contributing essentially to an identifiable common function. However, when such a function is not an end in itself but serves a larger function, the contributing systems are more appropriately referred to as subsystems of the larger system.

Where the functioning of a system is referred to, as distinct from the identifiable components of the system, that is, breathing rather than air, oxygen, lungs, diaphragm, trachea, or bronchi, we are speaking of a *process*, an operational manifestation of a system or subsystem. The operation of a system may be said to consist of a number of parallel and/or serial processes, or it may be said to result in a comprehensive or multistage process.

In the latter case, the system may be confused with the process, and vice versa, but there are significant distinctions. Among these is the fact that the process constitutes a function of the system, although the process may be inactive or that part of the system may not be currently functioning. An example would be the ever active scanning process of the Dew Line radars, while much of the rest of the system is poised but quiescent.

Another significant distinction is illustrated when a system functions, utilizing a typical process, but the result is not representative of the system's capabilities. Thus any one message from a communication system may not be consistent in content with the purpose of the system. As an example, the process is the same, regardless of the addressee and wording, whether the Polaris communication subsystem is used to send a birthday message to a crew member from his wife or a command from the president of the United States to launch nuclear warheads against an enemy.

Systems exist to provide and facilitate processes, but they add something to the order of effectiveness: they may make possible processes that would not be so otherwise, or they may combine processes, in parallel or in sequence, with synergistic effect. Thus the distinction should be observed between the system, which is characteristic of an arrangement of coordinately functioning elements, and the process or processes that it utilizes to effect its objectives and that have to do with a specific kind of purposive change.

In speaking of process, we should distinguish between processes aimed at physical change for inherently physical purposes and those primarily directed toward representing or facilitating symbolic change. Clerical processes involve necessary physical change through writing, typing, duplicating, card punching, recording on tape, printing out on paper, and the like, but the purpose is not primarily to produce these physical changes involved in the processes, but rather to utilize them to effect the recording of symbolic material and the changes in the values represented by the symbols recorded.

For the purpose of this discussion, a system will generally be considered to involve more than one process, and usually many processes or a comprehensive process. We will be concerned primarily with the effects of participation in such systems by individuals who are involved in only one of several or many processes in the system.

The characteristics of systems that distinguish them from simple mechanization appear to be related to size and scope and—usually—the interconnectedness of separate elements. Thus an operation carried on in separate but sequential steps does not become more systematic merely because one or many or all of the steps are mechanized, regardless of the separate gains in effectiveness. There must be some additional gain that is due to an advantage common to at least some of the steps before the stages of an operation may meaningfully be said to form a system.

Insight into the special significances of the terms *system* and *subsystem* may be gained by considering relatively new and only recently familiar applications. Thus the term *Polaris system* refers to all Polaris-type submarines plus all their essential supporting facilities ashore, as a unified operating whole. By speaking of a subsystem of the Polaris system, one does not intend to refer to a physically whole component, such as one submarine or one related shore establishment, but rather to one of the functionally interconnected operational characteristics pervading the whole system, such as the subsystems for propulsion, communication, navigation, targeting, and the like.

Indeed, in the family car, we find the ignition system, the fuel system, and the braking system, each functioning in parallel to the others through its separate but interconnected elements, each an indispensable subsystem contributing essential functions to the overall functioning of the car, which may be regarded as the family's transportation system.

When a system does not involve people internally (as the functioning of an automobile involves a driver who controls the subsystems but does not intervene in the internal functioning of the fuel, ignition, braking, and other component systems), it may be regarded as unitary; but it can still be very extensive. Thus a huge dragline or power shovel represents a system for the removal from a pit of earth of coal, iron ore, or other material in huge quantities, but it is operated and controlled by one man. It is made up of many subsystems—ground movement, steering, braking, lighting, heating, the systems for control of the bucket, the systems for control of power, and the like—and all are coordinated by a single operator. He is, of course, supported by bulldozer and other operations in the pit, by maintenance and repair activities, by systems for removal of the material extracted, by supply of electricity, fuel, lubricant, parts, cables, and other logistics, by financial, administrative, technical, and managerial functions.

A Boeing 747, similarly, represents a system made up of many systems and dependent upon systems outside the plane; yet all this interrelated functioning is under the sole control of the pilot.

Unified physical systems can provide tremendous extensions of the capability and productivity of individuals. The operator of modern steel mill equipment can turn out miles of accurately dimensioned strip per hour. A man at the control console of a large oil refinery or a huge electric-power-generating unit controls and maximizes the productivity of many millions of dollars' worth of capital goods. The primary dispatching executives of modern railroads, airlines, or bus, truck, or shipping companies determine the specific utilization of vast and widespread resources. The few in control of a fully automated factory are responsible for output that would have required hundreds or even thousands of workers only a few years ago. More and more extensive and elaborate physical systems have amply proved their value and are increasingly necessary in a modern economy.

And, increasingly, the systems approach is being applied to clerical work and to information systems utilizing electronic data processing; all sorts of equipment hardware are taking over clerical functions and creating new activities in our organizations—in effect, changes in organizations and new organizations.

It is inherent in the nature of systems that the whole is greater than the sum of its parts, and there is a tendency to subordinate to the interests of the system's effectiveness all lesser concerns as well as those that may not appear to affect it. This may be on the basis of short-range or long-range concerns.

Thus the safety of a scheduled passenger plane, as part of a system, is valued above the frustration and capture of a hijacker, because the fulfillment of the schedule of that particular plane is less important than the long-range consideration of minimizing danger to passengers. Conversely, railroad dispatchers have been known to sidetrack and delay passenger trains to give the right-of-way to freights, when freight revenue is more important to the system than passenger revenue. In the management of systems, the setting of such priorities is a major factor and can be a prime determinant of system effectiveness.

It is also inherent in the nature of systems, as of organizations, that there is a psychological distance between those who control the whole, or who are conspicuous focuses of system functioning, and those who serve separate, and therefore relatively minor, processes within the whole, and that this distance tends to be proportional to the magnitude and scope of the system.

Examples are the remoteness between the president of a telephone company and a switchboard operator or a repairman, or the gap between an astronaut and an hourly paid NASA employee. Even where there is much contact, mutual regard, cordiality, and interdependence, as between the commander of a bombing plane and his crew, there is also remoteness, which is enhanced by if not due to the complexity of the system and its importance to the functioning to which they all contribute.

The sociological effects suggested here can have profound influences on the motivations and attitudes of participants in systems, and thus on their performance. The significance of hierarchical factors may be greatly intensified in systems, with potentially serious effect. Chapter 6 will discuss this issue in greater detail.

As organizations grow in size, the number of echelons is likely to grow also, putting additional levels of management between the relatively few who can develop a broad, overall view of the organization, its objectives, and its functioning, and the vast majority whose horizons are limited by their localized and specialized circumstances.

The psychic distance between a senior manager and a low-level employee is increased by geographical factors, and multinational corporations particularly connote to their many thousands of employees, wherever situated, a sense of utter remoteness from the seats of power. The individual's feeling of relatedness to the organization is further attenuated by the multiplication and diversity of its activities, which force him to acknowledge that he is quite ignorant of many highly important interests and functions of his vast employer. The realization of aloofness in the employer-employee relationship is reinforced when the employee learns for the first time of some newsworthy development concerning his employer from the public press. This forces upon him the recognition not only that he is not an insider but also that he is no more entitled to news about his company than is the general public.

Such factors are becoming more and more prevalent, and characterize the employment of an ever-increasing proportion of the work force. The psychological reactions these conditions tend to engender provide a background of alienation and of generally negative utility, especially for the functioning of those operating within systems.

People Limit Systems

Economics, technology, and the nature of things all require that human beings participate in the functioning of systems. Even the most complex and fully automated systems require some human involvement, support, or intervention. Thus the behavior of people affects the performance of systems, and so anything that affects that behavior affects systems. A system designed to require more of people than they are likely to deliver is doomed in advance to less than its potential functioning, or to total failure.

Thus the human element should be seen as both a potentiating and a limiting factor. Whatever the system actually requires of human participation must be provided if the system is to function as designed.

Economics enters into the situation in several ways. Unless there is some other decisive factor involved, investment in the installation of a system is justified only as it provides an economic advantage—directly or indirectly—over an existing, less systematic operation. Then the investment in the system hardware is generally so substantial that its failure to function or its failure to function at the level of potential effectiveness may well represent an unacceptable economic loss.

On the other hand, the evolving requirements for human activity may be so exacting in their numbers or in the degree of competence required, or both, that payroll factors may increase the cost of systems operation unacceptably. And the cost of supplying additional hardware to reduce the human requirements may be economically unjustifiable.

Technology is necessarily involved, because the effective availability of hardware to replace humans and improve on tool-less human potentials is generally dependent both on the state of attainable technology and on the embodiment of such technology in economically practical form. Technology provides equipment that takes the place of or improves upon many operations normally or formerly performed by people with less mechanical reinforcement, or it makes new operations possible. But the specific requirements for the performance of such operations in any system may not economically justify the installation of such equipment.

For instance, if a normal day's work for a person at a particular task is represented by the figure 10, and a machine exists with a production capability equal to 1,000 for one day, the machine can equal the production of 100 workers. And if the system requires an output of 1,000, the machine would seem clearly justified. But if the system only requires an output of 10, it is probably better to keep the human on the job. Other factors and considerations will surely enter into the situation, and thus the critical point of choice should be determined according to overall system requirements.

Human beings are more flexible than machines, and for technological purposes may to some extent, and depending on capability, be regarded as multipurpose machines. But it is far easier to replace a human doing a simple, repetitive task than it is to replace a human doing a complex task. However, a human is also "open-ended," in the sense that he has capabilities for a greater range of activities than can be surely foreseen. He is more or less able to meet emergencies and to adapt to emerging, unexpected, or rare conditions. There is no guarantee that he will do this, or do it well. But it is generally quite predictable how a machine will act, and the predictability is certainly negative as to any reactive capability not specifically built in and programmed. Thus human participation in systems is sometimes provided for to make available some potential for coping with nonroutine events. The individuals involved may be trained and prepared for their responsibilities, and thus enhance the overall potential value of the system. But it is their actual performance that will determine system functioning when they are called upon to act as required.

If a person is needed and not present, ongoing operations dependent upon him must cease or go awry unless a substitute is available. If a person is busy on higher priority matters or so weary as to be less productive, the effect on performance will be preclusive or diminutive. Such truisms are as valid for individuals involved in systems as for others. Strikes, illnesses, and other unforeseen developments can affect the functioning of the system through the individual participant.

In all fairness it must be emphasized that a parallel dependence upon machines involves parallel risks. Machines fail or function improperly or inadequately, and such failures can cripple systems dependent upon them. However, the inanimate components of a system are generally responsive to programs of preventive maintenance, and when a failure occurs, immediate repair or replacement is usually possible. Humans are not as readily manageable in these matters, and a complex set of factors and considerations apply, differing with each individual.

It is in the nature of things that no system is infinite, and man-made systems have essential points of input and output, as well as points within the system where human participation is a factor. Errors of input will be processed by the system and emerge in the output ("garbage in, garbage out"), perhaps vastly proliferated. Similarly, a human error within the system's functioning will be embodied in the erroneous output. By nature, people cannot be eliminated from the environment within which the system functions, and from which it draws its input. Humans participating in the system's functioning cannot reasonably be expected to maintain a perfectly error-free, zero-defects level of performance at all times, with no exceptions.

Thus, people limit systems. That is to say, really, that people not only make systems possible but also, as participants in system functioning, constitute one of the limitations of systems capability and performance.

Systems Limit People

It is in the nature of systems that the performance of each identifiable component must meet specific, relatively rigid requirements or the whole system will be adversely affected. This is both a quantitative and a qualitative requirement.

Quantitatively, system productivity obviously is limited by the bottleneck effect when any phase of the system performs at a rate below that of the system generally, and especially when its rate falls below that of the preceding phase and/or the potential rates of succeeding phases. A typical example in a physical system would be a linear process, such as an integrated assembly line or the flow in a petroleum refinery or even a railroad right-of-way, where the output of one phase of the process would exceed the capacity of the next phase, thus requiring either a slowdown or storage of the excess. Typical examples in nonphysical systems would be a school system in which the teaching capacity for one grade would be significantly less than the output of the grade below, or a computerized information system with an output capacity below that of the input capacity.

Qualitatively, the relationship between the capacities of phases of systems can be far more complex. In the simplest cases, the effect of each phase is merely additive, and the overall result is cumulative. Thus in a system involving a simple mechanical assembly, the deficiencies in each part are accumulated by addition. However, additive quantities may have multiplied effects. Thus no crankshaft is perfectly balanced and no set of pistons is of exactly equal weight. The system allows some degree of tolerance—some measure of deviation from perfect exactitude. Ordinarily, these deviations will cause little trouble in the engines, and may even tend to balance out. But, in any one engine assembly, the heaviest piston in a set may be secured to the heaviest crank in a crankshaft. In such a case, two deviations are merely additive, but they will result in vibration that is a function of engine speed; and this magnifies the effect to a point that could conceivably be self-destructive.

In the nonsystematic assembly of an expensive luxury car or a special racing machine, skilled workmen trained to exacting standards can alleviate and perhaps virtually eliminate such probabilistic occurrences. But in the mass-production world of most cars, trucks, and tractors, economic factors demand systems of greater and greater technological pervasiveness and automative purity, and the quality safeguards needs must be built into the system itself. (One automobile manufacturer has advertised computer-matched pistons in a mass-produced car.)

In many physical systems a deviation from standard in one phase can doom the total of the system's production to unacceptable quality. This is especially true of food and chemical processes and some metallurgical processes. And in many physical systems the quality deficiencies originating in one phase are substantially increased or multiplied in successive phases. Thus a failure to wash adequately a raw material or product that is added to others may result in the dispersion of surface impurities throughout the whole mass of finished product.

The implications of this for the participation of humans in systems are especially significant in terms of the assigned tasks and responsibilities of the individuals involved. As physical processes become more and more systematized, the participation of most of the individuals involved becomes more and more circumscribed. As they become increasingly involved in the system, their performances are more specifically prescribed and their scope of discretionary activity is reduced, eventually to the vanishing point. At this stage, observation by them of nonstandard conditions would probably be at best a cue to summon the supervisor.

The employee nearest to the process in a physical system tends to be called on less and less for knowledge, skill, and judgment and more and more for consistent performance, at a uniform rate, of relatively simple and easily learned tasks. Often, in physical systems, he is a mere extension of the instrumentation. In nonphysical systems he is often called on for performance of repetitive tasks of childish simplicity. The economic and technological factors that foster this trend are easily understood, but some of the long-term effects are inadequately comprehended.

The employee in such a position not only has been deprived of incentive to develop a greater capacity to contribute or to improve his contribution to his assigned function but is forced to recognize that any improvement would be superfluous. His simple task has been designed to be sufficient as a phase or stage of the system and to be entirely adequately performed by relatively unskilled, inexperienced labor. The job description calls for certain limited activities, conducted in a specified way at a specified rate. There is no opportunity to contribute beyond these stipulations.

In fact, performance beyond that required is superfluous and in some cases would even have an adverse effect. Generally, the system allows some tolerance for inferior performance; conversely, in effect, there is usually little tolerance in the system for superior performance. In this sense, systems limit people. Systems tend to demand less and reject more of the individual's capabilities. He cannot, if he wishes, give more to the job than the system will receive.

If we compare the employee who tends a soldering machine on an electronics assembly line with the workmen who carefully assemble accurately balanced engines, we see major differences in quality of task with vital implications for the subjective reactions of the worker.

The skilled workmen balancing and assembling the engines are free—within rather broad limits—to utilize their equipment and to select from available components as their judgment may dictate. They are responsible for achieving a high degree of quality, and their only limit would be the unattainable, theoretical goal of total perfection. They are expected to determine significant characteristics of each component and to adapt their activities as the specifics of component characteristics require; and they are expected to adjust the time spent per assembly to the individual requirements of each assembly. Knowledge, skill, and experience are recognized attributes of the qualified worker, along with such characteristics as conscientiousness and reliability of performance; and an extensive apprenticeship is served to learn how to perform the functions involved on a journeyman level of competence and to earn employer confidence in the individual's capacity and reliability. Almost any degree of care, diligence, efficiency, and excellence they may devote to their work can find realization in the quality, as well as the quantity, of their output. In this respect, they are not limited. But it is precisely to replace such nonsystematic operations that many systems are developed.

These principles and generalizations about physical systems are even more significant in their application to nonphysical systems. An error in writing or typing, translated into a thousand copies, is propagated and may foster myriad adverse consequences. But when errors converge and combine with correct data, the scale of consequences may be vastly enlarged. Thus an error of a single digit in an item of computer input may load the computer memory with false data and so breed thousands of erroneously processed outputs.

The clerk who is familiar with a prototypical transaction is in a position to process all or most phases of clerical operations connected with it. Through familiarity with the subject he can appraise the data he handles, and he may note anomalies, disparities, inconsistencies, diversions from past practices, discrepancies, and many other indications of error. He has an unlimited opportunity to turn out work that is as complete, accurate, and valid as possible; and he has opportunity and incentive to learn many of the processes that contribute to or generate the content of his work.

But if this clerical operation is expanded and becomes part of a system, it will be partially mechanized and partially distributed along a flow line among a series of employees who are responsible for only a fraction of the overall operation. In such a situation each faces a limited task and has an extremely limited opportunity to contribute beyond the specific minimum requirements of that task. Whatever their aspirations or potentials, their opportunities to contribute to the operation are effectively limited.

The implications of this limiting of individuals by systems will be more fully discussed in chapter 6.

Physical Systems

Virtually all citizens of advanced countries have some degree of awareness of the many far-reaching and effective physical systems that serve the public. Our day-to-day and even our hour-to-hour experience involves use of and offers some degree of insight into the extent, complexity, and interrelatedness of physical systems and even of their information-handling components. It also provides some contact with their human components—the specialized, trained, behaviorally channeled personnel who answer our telephones and send out the bills, who operate our buses and subways, who sell us gasoline and police our streets, and who perform so many other services we consider essential. Every time we turn on electricity or gas we are bringing into our homes the products of systems that exist only to serve us.

And when we purchase an airline ticket and watch the clerk operate the computer-based, geographically far-reaching information subsystem that tells the status of reservation occupancy on any day of any flight from any airport to any destination within the system, we are—hopefully—observing excellent examples both of the specially adapted information subsystem serving the physical system, and of specialized, specially trained personnel acting in a prescribed, planned, predictable way, with their behavior patterned and channeled to meet the operational requirements of the system.

Even the personal attractiveness of the personnel, their uniforms, and the offices in which they work and their formula courtesy and helpfulness—are all to some extent due and responsive to the needs of the system and the plans and formulations of those who manage it.

Thus systems are very much a part of our lives—perhaps a major part, but surely a growing part—as we are users and consumers of their goods and services. They constitute important and influential elements in our physical and psychological environments. And if we earn our living as part of a system, then indeed we find our lives profoundly conditioned by the requirements the system imposes internally on those who help to make it function.

Modern industry, with its mass-production techniques and nth-generation machines, has evolved many physical systems and is expanding their scope and systematization with automatic controls, automation, and other technological advances.

Basic industries produce raw materials and convert them into usable industrial materials. In some of these, chemical processes are involved, which may focus on separation (as in the case of dissolving metals from their ores) or combination (as in the production of sulfuric acid). In others, electrical or electrolytic processes are involved, as in the production of chlorine or the refining of copper. In still others, physical separations take place, as in the treatment of magnetic ores, the milling of grain, or the centrifuging of whole milk. Physical combinations may be represented by anything from the alloying of metals to construction with reinforced concrete.

Industrial applications of heat range from the smelting of ores or the welding of metals to the distillation of petroleum and the use of steam turbines in thermoelectric power plants. Processes utilizing solvents are applied, for instance, in the treatment of cellulose or other compounds. Physical force is applied in such processes as drilling, blasting, and crushing ores or in the production of aggregates or road-building materials. And, of course, all source materials and their derivations must be transported from their place of origin, to and through each process, to the place of end use.

Any one of these processes may be developed technically in the direction of greater economy of unit production. This objective is usually approached in either or both of two ways: by the use of more advanced and larger capacity equipment, and by the division of the process into stages that allow for more efficient specialization of labor and equipment at each step.

These examples merely suggest a few of the essential processes applied in basic industries. Many of them are age-old. Their evolution into systems is relatively new. For examples, in Dutchess County, New York, there are a number of stone structures that were once furnaces for the reduction of local iron ores. These functioned in colonial times, and some survived to the beginning of this century, when Carnegie and the growth of Pittsburgh provided insuperable competition.

Men and boys would enter their woodlots on the neighboring hillsides to cut down and saw up the abundant oaks. Then they piled the wood, set it afire, and covered it with earth. Later they dug out the charcoal, loaded it in wagons, and hauled it with oxen to the nearest furnace, where they sold it. Other independent contractors used their own teams and rigs to haul in loads of iron ore or limestone at so much per load. Often they took it from their own lands and sold it to the owners of the furnaces. The owners of the furnaces had a few employees who charged and tended the furnaces and drew out the product. The crude iron was generally sold to blacksmiths, who used it to make nails, horseshoes, and crude agricultural implements.

The raw materials found in the natural environment were converted into usable form, but the various activities were carried on with only the loosest relationship to one another, based largely upon the voluntary independent action of the participants, loosely coordinated.

In a modern steel-producing company, there is still a rudimentary parallel to these primitive activities, in that raw materials are brought together and used to produce iron. But consider the differences, which suggest the nature of a physical system.

All phases of the operation are controlled centrally and are scheduled for optimum integration into the progressive phases of treatment and movement of the materials. Low-grade iron ore is dug from remote pits with huge shovels and draglines, conveyed to concentrating plants, pelletized, conveyed to ships, railroad cars, and barges, and transported to the mills. The ore is met there by coke from ovens owned by the company and fed with coal from mines owned by the company, and by crushed limestone from quarries owned by the company—all brought from different directions. The ingredients are combined in huge blast furnaces that run continuously, producing a constant stream of pig iron. (In some modern plants the molten iron is fed directly into an adjacent steel mill.)

The iron utilizes carbon, nickel, chromium, manganese, tungsten, and/or other predetermined and metered materials, in another appropriately designed and controlled furnace, to become steel. The steel is converted from ingot to billet to bar or rod or strip or other form or shape until it is ready for use.

All elements of all these operations are coordinated to produce maximum overall efficiency and economy. And it is precisely this capability for centralizing, planning, and effectuating controls for optimization of predetermined criteria that characterizes a system.

When Andrew Carnegie put together his steel company, he was introducing the systems concept into the new steel industry. And when J. P. Morgan formed the U.S. Steel Corporation, he was enlarging and extending the system to include the resources of the Mesabi Range at one end and effective control of the market at the other.

The making of hay is another example of the evolution of a system. The process goes back thousands of years, to the beginnings of agriculture. Today hay is still a major agricultural product. While the decline of the horse as an essential adjunct to transportation has greatly reduced one form of demand in advanced economies, the growing consumption of beef, veal, mutton, and milk products has greatly increased the overall demands for good quality, high protein hay. In addition, recently developed understanding of nutritional factors has led to an emphasis on characteristics of hay that would not have been generally understood a few generations ago. Botany, genetics, soil chemistry, plant nutrition, and other specialized sciences have offered new and superior species of grasses and legumes that produce hay with heavier yields per acre and greater nourishment and more value per pound.

For untold generations, hay was harvested by cutting it with knife, sickle, or scythe, letting it dry in the sun, and picking it up with wooden forks to be transported on the backs of humans or animals or in carts to storage in thatched stacks in the open or in the mows of barns. The labor was arduous and unremitting; and much of the hay was cut past its prime, or leached of its value by rain, or spoiled by mold. And the largest part of even the best hay was unnutritious fiber.

In modern haying operations, controls are exercised and equipment is utilized to maximize both the efficiency of the operation and the quality of the hay. Typically, the precise morning when the forage crop is to be of maximum value—a function of maturity and protein content—is determined. In one hay system, the forage is cut, pressed between rolls to facilitate drying, raked, and then put in windrows, all by one self-propelled machine at a rate of acres per hour. The windrowed hay may be baled by a machine that picks up the windrow, forms it into bales, ties the bales, and pitches them onto a truck or tractor-drawn wagon. Or the windrow may be picked up by a machine that chops and shreds the forage and blows it into a self-unloading truck or tractor-drawn wagon. The bales or the chopped hay are dumped onto conveyors that carry them to storage, where they are dried by heated air in a mow or in a plastic-covered stack. Or the self-unloading carrier may transport the chopped forage to a silo, where it is delivered automatically and preserved without drying in a near vacuum as "haylage," which can be conveyed automatically, when required, to the waiting cattle. Such hay systems, like so many other systems applications in modern agriculture, have given to large-scale, mechanized farming many of the significant characteristics that were once thought to belong exclusively to mass-production manufacturing.

It would seem obvious that to the individuals involved there are vast differences between the less and the more systematic of these operations. The old-time charcoal burner had little in common with the modern bituminous miner or the employee of the coke ovens. There is little similarity between the mining of magnetite with sledgehammer, hand drill, and black powder, and the operation of a huge dragline that can dig literally thousands of tons of ore a day from a tremendous open-pit. And there is little parallel in hand labor required per ton of iron produced between the old stone furnaces of colonial times and the modern continuous-process blast furnace. Similarly, the operator of large, efficient farm machinery has little in common with farmhands laboring with scythe, handrake, and hayfork.

Development of Physical Systems

However vast the difference that labor-saving machines may make to the individual participant, labor-saving machines alone do not constitute a system, though they may play an important or even indispensable part in making a system feasible. A physical system comes into being when the various stages necessary to bring about a planned transition from available material to desired product are functionally linked and interrelated in such a manner as to produce an overall gain in value derived per unit of cost, time, or effort expended, over that attainable by less systematic utilization of similar productive resources. A system provides a synergy derived from factors additional to the separate processes that it embodies. Unified control or management, superior coordination, integration of activities, and common supporting services are only some of the factors that may transform a set of processes into a system.

The efficiency and economy of the modern blast furnace is wholly dependent upon continuous operation that requires constant and unfailing supply of the essential materials. The system exists to actualize the potential efficiencies and economies by organizing and conducting an integrated continuity of activities; and the overall results are dependent upon the appropriate interfunctioning of each component of the system.

Similarly, the quality—and thus the value—of hay is affected by various conditions, and haying systems are designed to minimize exposure to adverse factors and to expedite storage at optimum value. No matter how well the separate operations may be carried out, the results will not be optimum if they are not carried out in proper relationship to one another.

Labor-saving machines provide increased productive capacity, but they are justified economically only where the increased production represents an economic gain. Similarly—and usually beyond this—systems generally find application when significant increases in quantity and/or improvement in quality are sought. Such objectives are conducive to the introduction of systems.

The physical systems discussed are related primarily to basic industries, generally those industries engaged in the taking of raw materials from nature and making them available to secondary or processing industries. Thus lumbering produces logs for wood or paper; oil wells feed the refineries and petrochemical plants; blast furnaces supply iron to be made into steel; fisheries send their catches to canneries; chemical plants combine primary ingredients to make plastics, fertilizers and other products.

In subsequent levels of industrial activity, such operations as fabrication, forming, shaping, and combining take place, followed by assembling and then finishing operations. Sometimes stages are combined; and in some cases some stages are not required. But, generally speaking, such stages are identifiable by the appearance of systems that operate within each stage. Some systems may cover two or even more stages; these are called linked systems. The fact that two stages are carried out in the same plant does not necessarily imply linked systems. For instance, Ford's River Rouge plant produces steel that finds its way into parts for Ford cars. If the steel goes continuously to casting or fabrication, so that there is a gain in efficiency, then the systems are linked. But if it goes to storage until required, the systems are separate. One objective of much modern industry is to maximize the advantages of vertical integration through the development of linked systems. With the proliferation of products and byproducts, an extremely complex linkage of systems can result; and if these proliferating systems are truly linked, substantial advantages may be attained.

Motivations for Employment

It is a distinguishing and obvious characteristic of industries that draw directly from nature that they must be physically located in advantageous relationship to the source of supply. Thus mines are necessarily located upon the lodes they exploit; lumber camps are placed in or beside the forests; oil wells are exploited where oil has been found. It follows inevitably that those who work in systems so located will have to endure the characteristics of the site, however remote from urban amenities, however adverse the climate, and however inhospitable the terrain.

This inescapable and obvious fact compels a process of self-selection that is generally quite effective. Those who cannot endure or who are averse to the conditions avoid them or leave them. Those who are indifferent to them or can tolerate them are those who go there and remain. It is only the exceptions who, for special reasons such as having no acceptable alternative, go to such locations against their desires or stay with reluctance. Thus the individuals employed in such primary industrial operations are, largely, those who adjust readily to or who find satisfaction in such environments.

The character of nature-exploitative industries generally is such that the immediate environment is adversely affected, and the work itself may involve proximity to loud noises, vibrations, or odors, or confinement underground, the necessity for protective clothing, the requirement for considerable exertion, or other factors not readily tolerated by most persons. Again, a self-selective process takes place, and those who have difficulty tolerating such conditions avoid them.

In the primary and extractive aspects of industry, quantitative rather than qualitative standards generally apply, and operational efficiency is likely to be measured grossly, as in tons per day. When production is readily measurable in relation to the efforts of identifiable individuals or teams, it lends itself to the application of performance incentives. Thus coal miners may be paid on the basis of tons per day brought from the face. Hard-rock miners sometimes earn high pay running tunnels on a dollars-per-foot contract basis. Lumberjacks may be paid on a board-foot basis. Fishing-boat crews share in the proceeds of the catch. Drilling crews are often on a performance bonus basis.

Incentive systems are a part of the appeal to the men—often highly skilled and experienced—who tolerate haphazard and often remote locations, lack of amenities, and inesthetic and exacting working conditions to bring the riches of nature to the industries that convert them to the uses of man. Such men are often fully absorbed in their work and adequately or even highly motivated by the judgment of their peers.

In the nature of such work, the participants usually understand the whys and wherefores of all they are called upon to do, and, generally, of what goes on in the environment. They are participants in a full sense: They not only play a part in the operation, but they usually understand what that part is and what relation it bears to the parts of others whom they know. They derive satisfaction from belonging to and being a part of the present, ongoing situation. Such men are often happiest when least constricted in what they do. As systems may constrain them, they often resent and resist.

In subsequent stages of industrial development, location is far less rigidly determined by geological and geographic considerations, and a far broader choice of environments exists, with the places of work preponderantly indoors and generally in or near cities or towns.

No matter what his place in such production, the individual worker is more likely to perform a rather narrow, highly repetitive section of the total labor contribution. Whether a skilled technician at a console watching instruments to control an unseen process or a semiskilled worker adding one or a few components somewhere on an assembly line, the individual is likely to feel (or become inured to) a sense of nonidentity, of squirrel-cage repetitiveness. The individual is, in effect, compelled to abjure the possibility of individualized contribution to the task and adjusts to the repetitiveness of the operation until a sort of automatism develops, and it becomes possible to maintain production while actually thinking of other things. Incentives do not usually apply; when they exist they are often so designed and administered that the sharing is not proportioned to the individual contribution but is on a faceless group basis.

Whatever is stultifying about such situations can only be intensified as the functions become absorbed into more and more comprehensive and demanding systems, leaving the individual less and less scope for initiative or even for accommodation to essential functioning. Under favorable conditions, a worker may meet daily production norms, for instance, by alternating intensive work with intervals of coasting. But, tied to a system, a steady, homeostatic rate of production is likely to be required. The worker is conscious of pressure, of being paced, of being dehumanized into a quasimachine.

Nonphysical Systems

In considering the significant distinctions between the two broadest categories of systems, we must have regard for the purpose or end result of the system. In a physical system, the purpose and end result are physical—a transformation of tangible entities, as in manufacturing, or the movement of tangibles from place to place, as with transportation systems.

Nonphysical systems are concerned with intangibles; but in this connection, that is a relative term, because of the limitations of the human sensory system in coping with intangibles externally. The human intelligence relates to the outside world through the sensory capabilities of the body, usually identified as sight, sound, touch, taste, and smell. Each of these senses deals with tangibles, and derives sensation from physically identifiable sources. Thus even the most intangible values are "sensed" through physical means.

It follows that a system dealing with intangibles must employ physical agencies. The intangibles are expressed symbolically in letters and/or numbers, but the symbols themselves must be detectable by physical means. They take the form of marks on paper or film, holes in cards or tape, and electrical charges on tape, discs, or cells. There is also physical movement as the physical embodiments of information are processed.

The true distinction between physical and nonphysical systems, therefore, lies in the functions to which these terms apply. Thus a physical system is concerned with physical results, a nonphysical system with nonphysical results. The two are, necessarily, parallel. As an example, consider a modern systematized film processing laboratory. The exposed negatives are delivered at the input end and processed by automatic equipment controlled by continuous feedback cycles, and completed positives emerge at the output end. This is a physical system. But such a statement ignores the purpose of the film—the very reason for the film's existence. Its content, represented physically by silver or other compounds deposited from the emulsion, is capable of informing, entertaining, or arousing emotions when displayed to people. These "intangible" effects represent the real purpose of the film, and thus indicate the significant content.

In many systems there is a combination of the two kinds, as subsystems, mutually supporting one another. The ultimate classification of a system rests upon its primary function or upon its specific function in a particular application.

Effects of Nonphysical Systems

Physical systems have had a profound effect upon our lives and enormously influence our environment; but it seems clear that systems and subsystems that deal with the intangibles of communication—systems that transmit, store, retrieve, manipulate, and transform data and information—are having an effect that is even more profound.

For the effects of these systems reach within our beings in the most direct way—to serve, extend, or challenge our intelligence. They do this, not merely by affecting the physical environment but by mastering space and time, by compiling knowledge and providing capabilities beyond human skills, and by putting within the power of a few the ability to perform tasks otherwise impossible to all mankind. They deal with the essence of those matters that distinguish the human from the animal, and they do much of it far better than any humans can.

More than machines or physical systems, the nonphysical system—often called the information system but perhaps more appropriately designated the organizational system—represents the most advanced imparting of hitherto exclusively human capabilities unto the inanimate world. Linked with physical systems, the nonphysical system represents man's most advanced realization of his creative capacity to form tools that enable him to extend his potentials closer toward their ultimate limitations, which seem now to be further away than ever.

All of us are affected by information systems, directly or indirectly. But those who serve these systems are especially affected.

The processing of information may be classified in a number of ways that offer general utility for discussion. One classification may be called the significant document class of information-handling process. This is the classical approach toward systematic processing of information, utilizing the original document, such as a letter, order, invoice, or bill of materials. In some cases, the documents are on a standard form, which simplifies processing. In other cases, additional information is added to the significant document itself—by machine, by stamp, or by hand—which provides a unified record and simplifies storage and handling.

Significant document systems utilize filing for storage. Where retrieval and use are required in several places, duplicate or multiple files must be maintained, involving the use of copies.

Alternatives to significant document systems require that relevant information be abstracted from the documents. The information may be transferred to a form or multiple forms, which lend themselves better to an information handling system, or it may be transferred directly onto cards, disks, tape, and the like for computer-based systems.

Computer-based systems store and retrieve information electronically, which provides flexibility and speed not otherwise attainable.

A single document can be filed in only one place. If it is necessary to refer to it under more than one heading, copies must be filed under the additional headings or the files must be cross-referenced to show the basic heading. Thus an order might be filed alphabetically under the name of the customer, with copies or cross-references filed under the headings of product ordered, name of salesman, and date of delivery.

The need for and advent of high-capacity hardware brought with it larger volumes of information/data to be processed and thus a need for clerks to transform raw data into processible form and to deal with machine outputs. This in turn led to the requirement for numbers of clerks to perform standardized tasks.

Standards for tasks to be performed make feasible the specialization of tasks. Specialization of task was a basic characteristic of the most primitive industrialization and was advanced to its prototypical, ultrarepetitive form with the advent of the assembly line, followed by the flow of paper in clerical operations. But specialization of task as an element in the division of labor in physical systems differs in two vital respects from specialization of task in clerical systems.

In a physical system, the characteristics of each unit and its progression from stage to stage are intended to be as nearly identical as possible. Limits of variance are specified. The effects of each operation are generally observable, as the performance of each task is generally observable, because each task involves physical operations essentially tangible and visible and real.

In a clerical system, each unit of work—each paper—is different, this being of the essence of symbolic items. Thus progression from stage to stage of each unit, while it may be parallel, cannot be identical. The items being processed involve infinite combinations of twenty-six alphabetical, ten numerical, and certain other symbols. Each unit of work, furthermore, bears within its own symbolic content the clues for its specific and individualized processing.

Information systems exist to collect, organize, summarize, select, abstract, manipulate, and otherwise process information in ways that increase its accessibility and usefulness. The characteristics of an information system are generally determined by consideration of two major factors: the nature of the information and the requirements of the system. The nature of the information determines the inputs; the requirements of the system determine the processing and the output.

So the simplest of clerical tasks requires, as a minimum: (1) correct perception of unique content through accurate reading of words, letters, and/or figures; (2) recognition and understanding of the appropriate application, in detail, of the generally prescribed procedure; followed by (3) actual effectuation of the prescribed procedure in appropriate and satisfactory form.

Extreme forms of specialization are becoming more and more common in extensive clerical systems, and the job content is often extremely narrow. But even the most minutely exacting of clerical tasks still requires some degree of performance in all three of the areas outlined. The direction in which this can lead is illustrated by the story of the Active Distributing Company.

Growth of a Typical System

In a typical distributor operation, the sales-oriented Active Distributing Company receives orders for its merchandise by letter, by telephone, and from its salesmen.

When the company was small, all orders were filed alphabetically in an orders file. As orders were filled and delivered, this was noted on the orders. Soon it was necessary for copies of the orders shipped to be filed also in the shipping department. As volume increased, a separate warehouse was established, and here also a file was maintained of copies of orders filled.

When bills were sent to customers, copies of the billings were filed with the orders in customer folders. As the business grew, copies of the billings were also filed alphabetically in a second file for bills receivable.

As collections became a problem, third copies of the delinquent billings were made and filed in order of due date. Eventually a fourth copy was sent to the salesman calling on the delinquent customer.

The treasurer and controller decided to keep closer check on receivables and asked that separate files be kept for billings that were one, two, three, or more months behind. This required monthly revision of each of the files of delinquent accounts.

At this point, when a delinquent account was paid, it was necessary to search the several delinquent files to find and remove the account's billings, remove the account from the receivables file, remove the order from the order file, and notify the salesman. The completed order and the paid billing, appropriately marked, were placed in a customer file.

Then the sales manager, who scrutinized all incoming orders, decided that he needed a new file with headings based on the items offered for sale, under which each order and customer would be listed. This required much of the time and effort of a new clerk in the sales department. Soon the dispersed data in this file proved unsatisfactory, and the sales manager had the file set up with a card for each customer, indexed under the item purchased. Because many customers purchased many items, this file became extensive.

The treasurer and controller then hired a credit manager, and he required customer files showing credit information and records of payment and delinquency.

Subsequently the purchasing agent decided that he needed a special file reflecting orders on hand so that he could keep active items in stock. Because pilfering was suspected, a set of special files was established to enable the controller to check orders delivered against goods received.

Then the new traffic department set up files to show locations of customers, bulk and weight of orders shipped, routings, carriers, and the like. A separate file was needed for insurance coverage of shipments, and other files were required for customer claims and claims against the insurers for damaged or lost shipments.

Meanwhile, of course, there were all sorts of books, ledgers, separate accountings, journals, records, ticklers, cashbooks, notebooks, private files, memos, instructions, directives, policies, and procedures.

And there were filing cases and manual typewriters and adding machines and filing cases and electric typewriters and billing machines and filing cases and calculators and bookkeeping machines and copiers and more filing cases.

The president and principal owner of the company called a meeting of his department heads.

"We sold 20 percent more merchandise last year than we did the year before," he announced. "And our average markup—the price we received over the cost of the goods to us—was up 2 percent. Salaries, commissions, and other costs of the field sales force were about the same percentage of gross as before. And collections were about the same percentage of gross sales. We could have had a very good year." He paused.

"We didn't! Office overhead was up 40 percent! That ate up all the extra profit we earned, and we didn't gain a thing! We hired a lot of new clerks, and most of them aren't half as good as the ones we had. And we're paying outrageous wages for unskilled people who don't give a damn. I want every one of you to start thinking seriously about how we can cut down on this clerical overload!

"Furthermore, you fellows are spending so much time at your desks now, it looks like our sales are beginning to fall off!"

But this brought little in the way of results. As the business grew, executives demanded daily, weekly and/or monthly summaries of the cash position, receipts, inventories on hand, receivables, past due collections, unfilled orders (dollar value and total of each item), shipments received (dollar value and total of each item), sales (dollar value and total of each item), shipments dispatched (dollar value and total of each item). From time to time demands were made for such data to be presented in percentages of totals or of bases, and even to be charted.

Additional staff was hired to conduct sales analyses, market research, pricing studies, and evaluation of results of advertising and promotion; all these accumulated more data and required more clerical assistance and produced more reports—and more employees.

Soon the requirements of personnel administration outgrew the capabilities of the office manager and the bookkeeping department, and a separate personnel department was created. This department had its own records of personnel, payroll, taxes, deductions, Social Security, employee benefits data, and the like.

Now the company had to rent more space to enlarge the offices and to store the expanding mass of files and records.

The president retained a paperwork consultant. This expert conducted a study and presented a report, for a fee equivalent to the annual salary of two clerks. He recommended a new multiple form with ten carbons, which simplified order processing and billing. This required several new electric typewriters and other additional equipment. There was no perceptible diminution of clerical load. In fact, the older and more experienced clerical employees were complaining about the overtime and about the ineptitude of the youngsters.

As the senior employees let their discontent be known, the executives began to be more and more aware of clerical errors in the records and reports coming to their notice. They also became aware of an increasing chorus of complaints from customers, generally arising from a mistake in filling an order, in billing, or in delivery.

When executives called in clerical supervisors to point out errors or to protest against the increasing incidence of error, they were met with a counterbarrage of complaints. The new, young clerks took no interest in their work; they were careless and disrespectful; and the supervisors and senior clerks were working overtime to handle the necessary corrections and to try to overcome or make up for the effects of errors. They wanted more—and better—help.

With the situation becoming more serious and profit-absorbing, the president called in several computer companies. Much executive time and thought were expended in deciding to rent—at a cost equal to the pay of more than ten clerks—a computer and associated equipment expected to solve the company's problems. Air-conditioned space was provided, several executives and employees were sent to receive special training, and card-punch operators were hired to convert all potentially live data into computer storage.

Most of the senior employees were quite unhappy—even disturbed—about this development. A few of the oldest asked for early retirement; some of those in their fifties and high forties and several of those in their thirties gave notice. The newer employees did not seem to mind especially. Their turnover was already high in any event.

A year later there was some new, high-priced help around: a manager of information systems, a manager of the computer center, and a few programmers. There were more card-punch operators. There were fewer older employees, but quite a few more younger ones. The executives were receiving more reports than they could digest. But they were busier than ever fighting the battle of paperwork.

And there were fewer employees now who knew how to correct errors or what to do when any nonroutine problem arose. The company was even short of people with enough experience to train new employees in routine jobs. As a result, customer complaints and inquiries were not handled as well or as quickly as before, and those not easily answered were backlogged on the desks of a few of the more competent clerks, who were in heavy overtime.

Every now and then an input error would spawn a

flood of derivative errors from the computer; and the resulting problems would scarcely be cleared up before a similar debacle would occur.

Meanwhile, many backlogged complaints and inquiries from customers were being taken up directly with members of management, as frustrated customers wrote or telephoned to executives to try to get action. Some grew frustrated enough to stop being customers.

It was impossible to dispense with the computer. But there was no diminution of problems. The harrassed executives blamed them on the younger employees, who performed inadequately and took no interest in their work.

The executives now received a stream of reports. But if, for some reason, any of them required that a new one be produced, he was often delayed for weeks or even months because there was a backlog in programming. However, the executives were now almost too busy to read reports; they were submerged in a sea of detail that flooded up from the clerical activities.

There was no effective reduction in the clerical force, but there were many changes. The general effect was to increase the number of inexperienced juniors and to diminish the resource of experienced seniors. New methods and procedures, designed around the computer, were aimed at minimizing the requirements for knowledge or skill. Somehow this failed to reduce the incidence of errors, even among the apparently more intelligent and potentially capable clerks. They appeared to take little or no interest in their expensively designed functions.

And once an error was made, it was likely to go through the clerical processes uncaught, until it ended in a letter to a supplier or customer, or in an order form or billing, or in a report, or in a punch card, and so into a computer output—or even into a program. And it seemed that when errors occurred and were propagated, there were more and more difficulties and longer and longer delays in rectifying the situation because there were fewer and fewer people who understood the system and its interrelations and workings, who knew the purposes of the operations and the implications of errors, and who knew what to do.

And these few competent people were busier and busier, and more and more backlogged, until the matters that had to be taken care of immediately—or finally—often ended on executive desks.

The harrassed executives of the Active Distributing Company believe that information systems are indispensable, but they earnestly wish there would be less difficulty and fewer problems in their use.

Many other companies have had experiences more or less parallel to those of the Active Distributing Company. Their troubles, generally, arose because of growth that greatly increased the requirements for paperwork and the simultaneous attempt to meet these requirements by systematization.

Unfortunately, the systematization created additional problems while not very satisfactorily meeting the needs for which it was introduced. The executives of the company were forced to spend too much of their time and energy on the secondary objective of trying to make the systems work better, rather than being freed to devote more time and energy to the primary objectives of their business.

The systems were adequately designed, and the equipment required was also adequate. The problem lay in the people who staffed the systems. They were quite unlike the veteran clerks they replaced—the ones who knew all about the product line and the accounts and the pricing and the salesmen's idiosyncrasies and all the other matters and items that enabled them to keep the company operating smoothly before it began to grow too fast.

No doubt better systems could have been designed, and systems could have been better equipped. But no such improvements would have greatly changed the problems upon which the company almost foundered, and these problems are typical of those increasingly encountered with the advance of systematization.

The executives of the Active Distributing Company are only too well aware that their problems and difficulties originate with the people in their systems, and some of them are constantly recalling the virtues and accomplishments of the former bookkeepers and clerks, and deploring the differences between those competent departed ones and the present lot.

But one day, over a two-martini lunch, the oldest executive struck a new note.

"Have you ever thought," he asked, "what it would have been like if we had asked those good old-timers of ours to do the kind of work we're asking the new crop of clerks to do? Put them at these same narrow jobs, and I'll bet in a few months, we'd be having pretty much the same troubles we have now!"

2 Evolution of Systems

THE opportunities for the advantageous installation of systems seem to correlate, in general, with the dimensions of the activities involved. Thus the greater the productive volume and the larger the number of persons involved, the greater the probability of advantage through a systems approach. This is true because the larger number of participants in the activity provides proportionate opportunities for specialization of component operations; and the larger volume of production provides the economic basis for the investment required, the effective utilization of specialization, and the possibilities for substantial economies and efficiencies.

Another common occasion for resort to systems is the recognition of potential advantage to be gained through speedy processing and availability of information. These can be paramount considerations.

Other requirements for the installation of effective systems include available technology, applicable managerial capabilities, supervisory experience, and adequate employees.

Growth of Systems in the Economy

In the larger economies, such as those of the industrially advanced nations, large-scale activities are numerous and proliferating, and they continue to offer expanding opportunities for the introduction of systems. In addition, as systems proliferate, there is growth in the development and availability of specialized equipment that makes systems operations simpler or more efficient, or makes new applications feasible. Also the pool of individuals experienced in various phases of systems operation is bound to grow, improving availability of qualified personnel for employment in new systems. Much of systems technology is general in its applications, so that advances in and more widespread knowledge of systems technology offer potential advantages to most systems.

In the more advanced economies, many systems are able to exist or come into being or grow and extend their services to support other systems. Thus basic communication networks—telephone, telegraph, and so forth—already exist to serve the needs of the public and are essential to the functioning of geographically dispersed organizations. In addition, the communication systems develop new services and facilities designed especially to service the communication subsystems of the larger organizations. Examples are the Data-Phone,* which provides links between elements of computerized information systems; the TWX, which provides links between remote writing and printing machines; and shortwave and microwave networks, which provide links between broadcasting stations or between information transmitting and receiving stations, such as mobile telephones.

* Service mark of American Telephone and Telegraph Company.

In addition, comparable technology has made available equipment that has made practical the performance of functions that create opportunities for new systems. Examples are the use of shortwave or microwave transmissions to portable sets, providing better dispatching and control of locomotives, trucks, taxicabs, agricultural equipment, harbor tugs, police patrol cars, and other dispersed elements of operating systems. Extensions of such systems functions are exemplified by the electronic scanner that "reads" the "signatures" of individual freight cars as they pass established checkpoints, and reports their locations to centers many miles away, thus enabling the railroad, in soliciting business from shippers offer to provide better freight handling with up-to-date information about the location of goods in transit.

Entrepreneurs have, of course, been active in entering the field of producing equipment for use in systems, which is now a major industry. And once they are committed to such a program—especially if they are actually producing the equipment—it becomes an essential element of their marketing activities to identify or develop potential applications for what they offer or can produce for currently recognized purposes or for purposes not yet in use; in systems or subsystems paralleling one or more already in existence and pioneering new but economically attractive functions. Improvement of the efficiency and extension of the functions of existing systems are also goals of suppliers, consultants, specialized staff personnel, and experts of all kinds.

The economic leverage of increased effectiveness through introduction of a successful system or through the improvement of an existing system can be substantial, especially in view of the cost of many installations, and the magnitude and cost of the operations they are expected to replace.

It is doubtful if many of the new developments in our economy would have been economically feasible without the development of advanced systems. For instance, the proliferation of credit cards and of many other extensions of credit-granting and installment payment practices now depend upon the functioning of highly developed computer-based systems.

The management processes of many large organizations depend upon the receipt of internally developed information transmitted and processed through extensive, specifically designed systems.

The deliberations of policy makers and the plans and decisions of executives of corporations are affected and influenced by information gathered and processed by and through other systems that exist to provide such services in government, the universities, foundations, and private enterprise.

Thus systems make new enterprises feasible and facilitate the fostering of profitability in ongoing businesses.

Growth of Systems in the Cultural Environment

The advent and spread of many larger and more extensive systems inevitably reaches out to affect those who are not directly involved internally in their workings. Interactions with most systems are on a basis that is both voluntary and inevitable. That is, the contact comes as a result of a private action—dialing a telephone call or making an airline reservation—but such contacts cannot be avoided if one lives a normal life, because the reach of systems is so nearly all-pervasive.

One of the overt symptoms of a culture's involvement in systems is the appearance of numerical symbols in conjunction with personal names—a distinction once accorded only to prisoners and the military. The first major personal numbering exercise affecting the population generally was the Social Security system; and for some time this was the only number affixed to the names of most Americans apart from their telephone numbers, the license numbers of their cars, and the numbers in their addresses.

But in the more recent past we have been identified in the systems of organizations we deal with through numerical designations (sometimes with letters attached). These include insurance policies; charge accounts with clubs, merchants, and restaurants; drivers' and other licenses; seats in planes, trains, theaters, and stadia, tickets for raffles, and lotteries; and address labels of magazines.

We also have account identification numbers in magnetic ink on our personal checks and in raised letters on plastic credit cards from general credit services, banks, oil companies, and the like where the actual physical form of the identification itself plays a functional role in the communication subsystem that records our charges and bills us as debtors of the system.

The onslaught of numbers has aroused resentment, antagonism, and resistance. Some of this was manifested when the telephone company moved to substitute numbers for the two or three letters that represented the subscriber's central office. Some mitigation of the burden of remembering or recording and relating to a variety of numbers has been gained by making some numbers serve more than one purpose: Thus the Social Security number is becoming a kind of universal numerical designation, because the federal government requires it on all records of payments to be reflected in income tax returns.

Numbering individuals has a depersonalizing effect because reference to an individual by a number seems to imply recognition only of the relation and significance of that individual to a system, and ignores all other considerations. This practice implies disregard of most individual characteristics, and thus negation of individuality. Some persons are more sensitive to these effects than others, but the effects are there, nonetheless, and they have an undeniable influence on our mores, habits of thought, and even our concepts of ourselves.

The cultural effects are felt more in some subcultures than in others. Those who receive or apply for public relief or are otherwise the subjects of social worker scrutiny are referred to by case numbers. After a time this tends to make the subject think of himself as a case, rather than as an individual with other claims to consideration. Thus identification of individuals as involved in systems can itself have a demoralizing effect. If this is reinforced by the impersonal behavior of those who administer the system and carry out its policies, this effect can be greatly emphasized.

Many large organizations utilize systems techniques and practices in the processing of people, including designation by number. A characteristic feature of such processing is the participation of a number of different persons, each administering a part of the processing. Examples are to be found in the employment offices of some major companies; the complaint departments of some department stores, banks, and other organizations; and the administration of claims by some insurance companies.

The individual who experiences such processing usually feels frustrated and helpless. Typical remarks are "There's no one you can talk to!" or "All you get is a run-around." The several individuals who conduct the serial processing are so involved in the system through their specialization and the limits on their functions, responsibility, and authority that they resist and even refuse involvement beyond the prescribed routine. Thus they become morally and psychologically conditioned through their participation within the system; and the individuals processed are conditioned through their contact with it.

The stultifying effects of extended participation in extremely specialized functions within physical and clerical systems have often been suggested, and even denounced. But the conditioning effects of outsider contact with the systems impersonality of large organizations have been largely overlooked. This has aroused indignation, but the result has generally been the expression of targeted resentment rather than the recognition of a broad pattern of cultural effects.

We are learning to live with systems as they are. And we are likely to accept their occasional annoyances, as we accept their indispensable and often enlarging benefits.

There are demonstrated differences between attitudes, values, personal characteristics, and behavior developed in individual activity and those developed through participation in group activities. The man who chooses to work for himself is a different kind of person from the man who seeks employment. There may be some question as to which is cause and which is effect, but the longer they follow these diverse courses, the more different they become.

Such differences have significant cultural effects, and our culture exerts potent influences on the individual in the direction of group participation. Even professionals—doctors, lawyers, architects, accountants, experts and consultants of all kinds—who were the individual practitioners of a generation or two ago are moving toward group practice, more intensive affiliations, larger firms, and more active associations. Many with such professional qualifications are now employees of corporations.

Systems offer many actual and potential advantages that tend to foster the larger group activities. Within the system, there are usually requirements for uniformity, conformity, subordination, and compliance. Thus the trends to bigness and away from individuality of contribution are encouraged, made more feasible, and supported by the proliferation of systems and of their applications. The cultural effects are widespread, profound, and powerful. And little is done to detect or measure, much less limit or control, the results.

Systems make some important direct contributions to cultural activities through improvement of related services. Examples are the facilitation of library services and the storage and retrieval of data for research projects, as well as the fostering of activities that gather information and process it to make available otherwise unattainable results. Even the columns of many daily newspapers are enriched with background information supplied by systems developed for such purposes.

Growth of Systems in the Governmental Environment

Because government touches everyone at one time or another, the extension of systems into government operations is far-reaching. This is true on the levels of federal, state, and local government. The military provides the truly outstanding example, but it may be enlightening to consider others that are less conspicuous.

The Bureau of the Census systematically collects and compiles a wide variety of information about all citizens, which can be treated statistically to provide information keyed to such indexes as geographical location, income levels, age brackets, sex, family size, educational levels, and others. The information derived from such data is available to all and makes possible a relatively systematic approach to great masses of the population, as in the planning of marketing campaigns, or in the location of factories, warehouses, stores, or other physical elements for organizational activities.

The specialized activities of the census, covering such areas as farming and manufacturing, probe even more deeply and provide a wealth of information that is the basis for many kinds of plans and projects for the widest variety of purposes. This offers the possibility of ever more predictable results from the operation of systems that relate to these significant factors.

The involuntary conscription of young men into the armed forces is conducted through a specially adapted, rather loose civilian quasisystem. The Post Office represents a huge (purported) system, struggling under too many handicaps. The administration of much of the legislation passed by Congress requires the creation of extensive systems. Social Security, the Federal Bureau of Investigation, and the U.S. Weather Bureau are only a few of many impressive examples of the systems created to give effect to policies established by the government.

The federal government's system of physical facilities

such as beacons makes possible the safe use of the airways by the planes of air transportation systems. This is reinforced by federal programs that help finance airports, ground-control equipment, and the like. Many systems depend upon the postal service, and could not exist without it. The federal highway system is a major factor in the functioning of many private and public transportation systems, as is the federal system of aids to marine navigation. The central files of the Federal Bureau of Investigation constitute an essential resource for the functioning of state and local police systems. The Library of Congress system of indexing facilitates the operation of thousands of public and private libraries around the country. The weather-reporting system renders invaluable service to the widest public, and especially to agriculture, airlines, and other activities dependent upon climatic conditions.

In addition, the many regulatory bodies—Interstate Commerce Commission, Federal Aviation Agency, Federal Communications Commission, Food and Drug Administration, Federal Trade Commission, Securities and Exchange Commission, and others—intervene in and affect the operations of many nongovernmental systems by specific regulations. Major government departments also have profound effects on nongovernmental systems. For instance, the Department of Labor can affect the conditions of work and the remuneration of employees. The Treasury Department and the Bureau of the Budget exert potent influences through taxes, customs duties, and fiscal policies that affect the economic climate. Health, Education, and Welfare reaches with decisive effect into many systems, both those of local governments and private ones. Thus the federal government represents many factors that foster the growth or control the development of systems outside the government.

States operate highway systems, special systems of communication (as for the state police), systems for the assessment and collection of taxes, and many others. States also provide for the administration of justice through court systems, and they have responsibility for the administration of the educational systems. (Some states are further advanced in the development and application of systems than others, and some states tend to exercise their powers centrally, while others tend to delegate more powers to local governments.)

The state governments operate systems to serve a number of their special functions, such as registration of automobiles, trucks, buses, trailers, boats, boilers, and, varying from state to state, other equipment such as X-ray machines, dance halls, gambling facilities, and so on. They also license a wide variety of activities—from ordinary drivers, truck drivers, chauffeurs, and learners to professionals such as doctors, dentists, lawyers, engineers, and others, including steam engine operators, chiropractors, masseurs, casino croupiers, handlers of explosives, and even dog owners. In addition, states issue permits to operate all sorts of specialized businesses such as distilleries, bars, restaurants, food manufacturing plants, liquor dispensing outlets, and the like. States exercise supervision over many aspects of industrial activity: safety, minimum wage levels, fair labor practices, and compensation insurance. Many of the administration requirements are met through the operation of more or less well-adapted systems.

Local governments vary tremendously in population,

geographical size and character, economic and industrial levels and characteristics, and in the extent of state powers delegated to them. Cities, in general, operate educational systems and street and highway systems. Other local functions, such as police and fire departments and garbage disposal, may be more or less systematized and usually include subsystems. Water supply and sewage disposal facilities are essentially and physically unitary in character, and thus are logical foundations for systems.

In addition, local government and the effects of its functioning impinge daily on the lives of the inhabitants and the conduct of business at many points. This necessarily affects the feasibility and functioning of nongovernmental systems. Thus delivery systems using the streets are affected by traffic control. The distribution of electricity, gas, and steam by utilities, whether privately or publicly owned, is subject to the activities and regulations of local government. The educational system will have a long-term effect on the capabilities of employees in the locality, which in turn may affect industrial and clerical productivity, and thus the functioning of systems in the area. The apparent efficiency of police and fire departments has an effect on insurance rates. Local policies on taxation can exert a profound effect on real estate values; local planning and zoning can control land usage; local policy on licenses and permits can affect not only individual livelihoods but also the operations of home offices, franchisers, suppliers, and other interests—often involving systems, and often outside the locality.

However, one of the most extensive effects of the operations of government at all levels has been the fostering of the bureaucrat, the so-called typical government employee, who is often called a product of the system. While the disparagement implied by this kind of classification is undoubtedly unfairly applied to many well-intentioned, hard-working, and service-oriented employees of various governmental agencies, it is nevertheless true that huge government bureaus have provided the prototype of the clerk who "goes through the motions" with no real concern for the content, value, or effect of what he is doing.

To the extent that this symbol of bureaucracy may have validity, it represents one of the basic effects upon individuals that can result from participation in systems.

Origins of Systems

For several millennia of recorded history, man-made systems were relatively simple, and their administrative requirements were readily recognizable (if not met). They were, necessarily, based upon available technology, which was in a primitive state of development.

Consider the problems of a Caesar, an Alexander, or even a Napoleon, seeking to transmit a message to the men in his armies. We have heard many romantic tales of how such great leaders inspired their followers with passionate and heroic speeches. But consider the facts. The broadcast, the public address system, and the bullhorn had not yet been invented. The greatest of generals had to rely on the carrying power of his own voice, and only a crowded, huddled fraction of his forces could hear him. Furthermore, almost all the soldiers were illiterate, and so could not read any messages intended for them. The great leaders had to adapt to such conditions, as well as to administering widespread conquests through communication systems entirely dependent on the horse-borne courier.

Industry was in the hands of the master workman, the journeyman, and the apprentice, who functioned in small workshops, usually in the homes of the guild members. Hand tools were the major equipment. Government was largely authoritarian, which provided the simplest and (under the circumstances) most effective system for controlling the population and the economy.

Leadership of all phases of government and the military was drawn from a hereditary elite that clustered around the absolute monarch. There was little or no protection in law or practice for the rights of the individual, which, in any case, were imperfectly or not at all recognized.

The church presented the outstanding example of a large organization; it was international and capable of functioning effectively in relation to a variety of nonspiritual objectives. It was no coincidence that priests, prelates, and other members of the clergy were among the few who could read, and it could be argued that this fact alone made it feasible for the pope's system not only to function as it did, but even to provide important support and assistance to lay governments.

The creation of capital goods in the form of fortresses and cathedrals enabled the holders to attract and dominate increasing numbers of others. They could extend their ownership or control of land, which was the primary means of production, then command the loyalty of those to whom they granted it and exploit the labor of the wholly controlled serfs who worked it—the feudal system.

The advent of a new kind of capital goods in the form of artillery not only devalued the fortresses but enabled the holders of cannon to attract and dominate even more extensively. Thus the developing royal houses gained ascendancy over their most powerful nobles.

The support of an artillery train and the conduct of a bombardment required an extension of technology, organizational complexity, and operational control beyond that known to the Romans and other ancients, with their catapults, siege towers, battering rams, and other crude implements. The military leader was less and less the dashing hero who led the assault and more and more the tactician who planned and conducted it.

With more powerful weapons and with the larger forces that improved organizational techniques made feasible, the command of armies gradually became more a profession and less a hereditary privilege. (At least, there was increased recognition that not all those hereditarily eligible were equally competent to meet the more exacting requirements for higher military command.) Paralleling this development, the domination of greater areas and more diverse populations demanded an enlarged capability for administration; and governmental efforts took more systematic forms.

The Industrial Revolution and the Worker

With the advent of the Industrial Revolution, manufacturing processes began to develop beyond the traditional elementary stages; and with the advent of the steam engine and the factory the worker-owner relationship became largely depersonalized and dominated by the need to serve the machines and keep pace with them.

The huge surplus of available labor nullified the bargaining power of those seeking employment, and in any case there was no tradition in societies emerging from feudalism of serious regard for the individual—especially if he was of the lower classes. The need to work in order to eat dominated the capital-labor relationship in the industrially developing nations until other factors intervened—notably the growth of labor unions and, in the United States, the availability of alternatives in the frontier.

Nevertheless, the technology of mass production, based essentially on specialization of function, division of labor, and interchangeability of parts, changed vastly in the few generations from Eli Whitney to Henry Ford. Whitney replaced the gunsmith, who made one musket at a time, by employing a group of skilled craftsmen to use specially created tools and machines to produce the interchangeable parts from which muskets were assembled. Ford originated the assembly line that required very little of its workers in the way of skill, knowledge, or experience, but demanded so much in physical stamina and in tolerance of the monotony of repetitive operations that few men endured it long, despite precedent-setting wage levels.

The Ford assembly line was not only a milestone on the road of industrial productivity but it also marked a turning point in the philosophy of wages. Everywhere, the untold thousands of poor unfortunates (many, recent immigrants) toiled to the limits of their endurance—often at intolerable tasks—for a pittance that hardly supported family or life itself. They had little choice; the alternative for many was starvation. But in Dearborn, Ford offered a premium wage for acceptance of ordeal conditions if the recipient could and would perform as required.

Ford had to modify the exactions of working on his lines, but in essence, the constructive primary principle he recognized and espoused is now dominating or tending to dominate the conditions of employment throughout the United States and in the industrially advanced nations, introducing dynamics scarcely imagined in but still traceable to the days of the Model A. Social and humanistic considerations, of course, have advanced greatly since that era.

Dominant among many factors now is the advent of systematization of clerical as well as industrial processes and operations, so that the requirements for individual performance in each job or position tend to be rigidly specified, controlled, and limited.

Full or near-full employment and new social legislation, which mitigates the economic effects of unemployment, are now basic, underlying factors in freeing individuals from unacceptable conditions of employment.

The result of such partially opposed forces is identical with Henry Ford's original pioneering solution—higher wages.

It is true that the conditions of work in the physical systems have improved greatly. The formation of the United Auto Workers and other unions of the so-called industrial pattern (typical of the Congress of Industrial Organizations, in contrast with the craft unions of the American Federation of Labor) gave great bargaining power to the unskilled and hitherto unorganized workers and enabled them to demand and receive great improvements in wages and working conditions.

At the same time, educational levels and standards of living have gone up; workers have learned to utilize and enjoy greater leisure; available entertainment and recreation have been vastly elaborated. Among the results have been far less tolerance of boredom, enhanced self-concepts, and higher evaluation of the subjective aspects and qualities of work.

This critical development has led to an increasing demand that work be satisfying or, at least, not frustrating or stultifying, interesting rather than dull, expanding the capabilities rather than bypassing them, gratifying rather than destructive to the ego of the employee.

The managers of physical systems are line managers; the manufacture of products is the primary function of their enterprises. The cost of labor is an important but often not a determining factor in their policy formulations, and they usually adapt or adjust other factors such as tooling, design, marketing, and pricing, in order to pay the wages necessary to attract the required labor force. An example of the typical results is shown in a news item: "Auto industry pleas for Government help in training mechanics were rebuffed publicly today by a Labor Department official, who said the problem was wage rates that paid mechanics less than unskilled men who bolted bumpers on new Fords."[1]

In effect, the economic leverage of the system allows a pay level high enough even to attract skilled mechanics to repetitive, unskilled tasks, limiting their performance to levels well below their competence. Over the years, systems have exerted such attraction over millions of workers with hard-won and valuable skills in less rewarding fields, such as milking cows, plowing fields, blacksmithing, and other activities that have become mechanized or even industrialized in the last few decades. But the workers making the transition from the exercise of

[1] *The New York Times*, March 18, 1970.

personal skill to participation in the mass routine of a system have also had to make psychological and sociological adjustments involving varying degrees of frustration, which are readily accepted by some and all but intolerable to others.

Systems and the Worker

Systems are relatively inflexible; it is incumbent upon the individual to adjust. The trend, unfortunately, is toward making personal adjustment more rather than less difficult. It is essential that this trend be reversed.

As nonphysical systems develop, they parallel many of the characteristics and effects of physical systems. But at least two basic distinctions are relevant here.

1. The nonphysical system is seldom regarded as a line activity. However vital to the conduct of the enterprise, it is usually related to staff functions, whether on corporate or departmental staff. The evaluation of the function and considerations bearing upon this are necessarily considered in relation to the line functions they serve, rather than in a direct relation to the making of a profit. When the staff function is seen as directly related to profit, it is usually because of facilitation of a line function or as the result of a saving.
2. It is generally impossible to apply useful quantitative measures to the performance of individuals in nonphysical systems (unless they are managed on a quasi-industrial basis). This contrasts with the situation typical of physical systems, where the task of each individual is specifically quantified and individual performance is or can be measured usefully.

The general effect of these two distinctions between

physical and nonphysical systems is twofold: Unskilled participants in nonphysical systems are generally paid less than those in physical systems, and they are generally less productive.

These conditions are, of course, widely recognized, and efforts are made to overcome at least the lower levels of productivity. The major approach is in the direction of greater industrialization—increasing mechanization and automation, and introducing industrial engineering principles of work flow, rigid procedures, and methods based on work simplification. Work standards are introduced along with detailed job descriptions, and job-related rating is emphasized.

The drawback to all these measures is that there is no way to attach the controls to the individual. Lacking motivation and interest in work results, attention levels fall and quality of work suffers. To stretch a simile, it is like trying to drive a car when the steering wheel has little effect on the front wheels, the accelerator is out of reach most of the time, and the brakes sometimes go on when least wanted.

There is another fundamental consideration that is often overlooked in the management of nonphysical systems, but which requires adequate appreciation if some of the most important factors governing the functioning of the people in nonphysical systems are to be understood. This has to do with the critical, basic differences between working with symbolic materials and working with tangible things.

The difference between many physical tasks and most clerical tasks derives from the nature of symbolic activity, as compared with the manipulation of objects, and is measured in the kind and degree of attention required.

In assembly and processing operations and in working with production machines generally, the emphasis lies in sameness and repetition, in the essential similarity of the successive operations. Thus the functions of eyes and hands can be fixed in a pattern, and the attention can be trained to near-automatic functioning, to be triggered into adaptive activity only through awareness of a deviation from the prescribed routine—when something is not the same.

In clerical tasks, however, the worker must deal with symbolic content, not with the reality of substances and things, and the whole thrust of the function rests upon distinguishing the *differences* in appearance and in significance between one symbol and another, and between one piece of paper and another. Even the simplest of clerical tasks, such as sorting mail or key-punching computer cards, demands consistently accurate interpretations of symbolic material, plus specifically cued action based upon those interpretations. This calls for a level of attention and alertness of a quite different order from that required for repetitive mechanical tasks.

The requirement for sustained concentration on what are often to the worker essentially meaningless tasks is generally typical of clerical systems, just as the requirement for essentially repetitive manipulation is generally typical of physical systems involving mass production.

3 Advantages of Systems

ONE of the primary advantages inherent in systems is illustrated by the telephone system. One telephone by itself is useless. Connected with another, it forms a minimum circuit. For each telephone added, there is an increase in the potential advantage.

The instrument by itself has no utility. It is only in relation to others that utility is created. Thus the utility is inherent in the system itself. The connection of individual telephones with exchanges and the interconnection of exchanges into centrals and so on up to the most widespread networks are all technological exploitations of systems potentials. It is only the incorporation of each telephone into the system that accomplishes utility—and justifies the telephone.

A parallel example is the reporting and predicting of weather conditions—a service vital to many, from farmers to airlines. A single observatory, however well equipped and expertly manned, can only report the past and note the present local conditions with any degree of competence. When it comes to prediction beyond the short-range, it is virtually helpless. An observer upwind can be helpful; and a full-fledged observatory that happens to be upwind at any given moment could provide the basis for a relatively reliable prediction as to the impending weather that comes from that direction. Thus a planned network of appropriately located observation points, mutually reporting to one another, can each serve one another.

But this potential is truly systematized when all the available information is collected, recorded, integrated, and coordinated at a central point, and then disseminated for utilization locally. Thus the positive advantages of greatly improved predictability of weather conditions are inherent in the total system, rather than in the individual local observatory and its staff.

Inherent and Potential Advantages of Systems

Systems can be made compatible with one another, enabling them to interact and extend each other's effectiveness. Uniformity in critical elements of design and similarity of material (whether substances or languages) with which the systems are intended to deal make feasible many useful or potentially useful forms of coordination or cooperative functioning among systems. Thus computer programs of common applicability can be utilized on similar equipment; a car of one railroad can run on the tracks of another railroad of the same gauge; most separately owned telephone companies can interchange calls. Such standardization leads to great extensions of the usefulness of some classes of systems, from international aids for marine and air navigation to the voltage and amperage for domestic appliances.

The growth and development of systems of all kinds have led to recognition of the present and potential advantages of functional compatibility of systems, subsystems, components, and even methods and means for managing systems. Thus a component or a unit of peripheral equipment developed for one application or kind of application soon finds other applications; and techniques, programs, procedures, methods, and other elements of systematized operation are ever more broadly utilized as they find application elsewhere.

Sometimes knowledge, insights, or techniques gained in the development of a system in one field may prove helpful or even applicable in one or more quite different fields; and this may provide for compatibility between or among the systems. The development of containerization for easy interconnection among ships, trucks, and railroads thus greatly facilitated transfers between transportation systems.

Over the years, an implicit convention has led to the widespread use of letter paper in sheets measuring 8½ by 11 inches. This simple concurrence has made feasible the mass production of an enormously wide range of products from envelopes and typewriters to filing cases and duplicating equipment—with all their elaborate or exotic developments and imaginative peripherals.

This combination of the proliferation of systems with improvements in technology and know-how and the increasing interchangeability of elements (and thus the interconnectability of systems) has resulted in an exponential growth of possibilities for new extension of systems applications and a consequent opening up of apparently unlimited potentials.

The potential advantages of systems are beyond imagination. Some possibilities are suggested, of course, by past achievements. The Manhattan Project that developed the nuclear fission bomb during World War II constituted an unprecedented and largely systematic organized effort, with virtually unlimited claims on available resources (human and inanimate), to achieve not only a major scientific breakthrough but also a practical end-use product, and a substantial production capability, utilizing what had been only experimental processes. Then the postwar development of the Polaris system, with nuclear submarines and nuclear-tipped long-range rockets, demonstrated the conduct of an even more systematic large-scale project, similarly involving many component objectives; and this made even more substantial contributions to the techniques for systematic management of complex enterprises. Among these was the Program Evaluation and Review Technique (PERT), which, following the Critical Path Method, has become a useful and widely applied tool for project management. At about the same time, the vast, systematic attack upon poliomyelitis culminated in the development, testing, large-scale production, and near-universal administration of the Salk and Sabin vaccines, and thus to the near elimination of a dread disease.

The nature, scope, and achievements of such projects have done much to create faith in the extraordinary capability of man to accomplish near miracles through the operation of systems. And the moon landings, orchestrated by the National Aeronautics and Space Administration, dramatized the all but fantastic potentials of such large-scale projects and made obvious the fact that highly developed systems—and systems of systems—were indispensable to the accomplishment of such project objectives and alone made them possible of achievement.

"All systems go!" became the appropriate watchword of the Age of Systems.

In our daily lives the values and effects of systems are so basic and familiar that we are usually not very conscious of them until they are withdrawn. Labor disputes and major equipment failures occasionally deprive us of amenities we normally take for granted—a running water supply, functioning drains, electricity, gas, hot water, room heat or cooling at a touch, television and radio reception, telephone service, garbage removal, subways and buses, charge accounts and credit cards, chain and franchise operations that supply us with food and clothing, daily newspapers, police with walkie-talkies or in radio-dispatched patrol cars, mail delivery and pickup, and so forth *ad infinitum*. And we come within the scope of many, many more systems when we go to work.

Every day new services are offered that utilize existing systems and may make them even more useful. Every day new products are offered that somehow fit into and supplement or extend existing systems. Some of these services are systems in themselves, and some of these products are produced by or even offer new systems.

Systems beget systems. Many systems seem also to create the opportunity for alternative or add-on features, and some of these make possible other systems. We are in an age of exponential proliferation of systems, and our lives are more and more involved with and dependent upon them.

All this surely will go much further, but this prospect raises many questions.

Advantages of Systems to Organizations

The size and complexity of many modern organizations have both cause and effect relationships with the existence of systems, because it is the broader scope of organizational functions that makes extensive systematization desirable or necessary. It is the extent of organizational resources that makes such systematization feasible; and it is the potential of systematization that induces men to extend functions and enlarge organizations.

Organizations grow in several dimensions. Horizontal growth may be due simply to increasing magnitude of operations—more people required for the same operations. While this is a relatively simple situation, it opens the door to improvements and extensions of physical systems, and it creates or enhances needs for organizational functions that may require or justify system solutions.

Horizontal growth usually requires organizational adjustment in the form of vertical or hierarchical growth. This tends to diminish the contact of the leadership at the highest echelons with the operating levels below. Information systems can bridge the gap at least partially with available factual data and can process the available information to produce more assimilable summaries or abstracts.

Horizontal growth may also be due to the addition of new operations. The addition of new operations involves an increase in complexity and of actual or potential internal interrelationships, which is more than additive and may be exponential. This requires careful organizational adaptation and can place great strains on planning and control as well as on direction. The introduction or extension of nonphysical subsystems to facilitate management often follows.

In a typical corporation of substantial dimensions the factors that make for complexity are obvious. There are hundreds or even thousands of employees, usually organized into operating divisions and subsidiaries. Each of these has its functional departments, generally tied in various ways to corresponding staff departments at the corporate level. Each department has its own operating responsibilities, with differentiated functions assigned to the personnel. Geographical dispersion and the number of vertical echelons vastly complicate the situation.

Executives and professionals concentrate on integral operations vertically in the organization (departmental) or on certain similar but dispersed activities, wherever found (staff). Such functions concatenate organizationally and can serve to focus attention on or summarize information concerning identifiable subjects. But often this routine functioning does not provide the upper levels of management with the current information necessary to conduct the affairs of the organization effectively.

The problems of information flow may be usefully considered in terms of three dimensions—quantity, complexity, and time. Organizations constantly generate vast quantities of data, and this must be processed, digested, and summarized so that it becomes available to management in a useful and suitable form. Organizations generate many different kinds of data on a broad variety of subjects. Each of these must be handled as specifically required; and, of course, a vast array of relationships among these separate lines of data may have real significance, thus creating the complexity. And some information should be acted upon regularly or with minimum delay, creating the time factor and the element of urgency.

Generally, the various processes involved—physical and nonphysical—are implemented at first on a nonsystematic basis or with a minor degree of systematization. This may be quite satisfactory for small units. As the organization grows, the processes are linked and systems are developed. Eventually—especially with the advent of automation and the computer—control is centralized, networks are formed, and the systems are linked. In many situations, this development appears to be essential for the effective management of large and diversified organizations.

Once the transition is made to centralized control of a comprehensive system, the facilities and capabilities usually exist to extend the ramifications and applications of the system for control to other areas of the organization, to develop additional information, and to expedite the availability of information to geographically and hierarchically remote points.

The high-level executive, sitting in the executive suite in a major city, can review on daily and weekly bases a few sheets of paper that represent the vast quantities and varieties of data generated in many locations through a wide variety of operations and activities on many levels. The executive tends to become quite dependent upon these information systems outputs, which provide him with some degree of awareness of a number of ongoing activities and may alert him to situations requiring his attention.

The value and effectiveness of an information system may depend largely or entirely on the degree of adequacy and speed with which it provides relevant and sufficient information in appropriate form to those members of the organization responsible for acting upon it. A well-designed, properly functioning system can be an invaluable as well as an indispensable adjunct to successful management.

Knowledge is power and the possession of relevant information puts executives in a position to act with greater likelihood of success. But by the same token, it increases the gap between those who are in a position to know and those who are not. This gap may have serious consequences. The system adduces the information, but the determination of distribution is an organizational matter—once the facilities exist.

Many organizations that have demonstrated the capability to make information available have exercised questionable judgment in utilizing that information and/or in disseminating it. The system that provides the organizational leadership with information does not also provide the leaders with judgment as to its uses or applications, unfortunately.

Dependence upon systems is common in organizational management and often inevitable. The wisdom and effects of such dependence are discussed in this chapter and other factors in the relation between systems and organizations are discussed in Chapter 5.

Advantages of Systems to the Economy

Another basic advantage of systems lies in their economic effects: They render economically feasible the use of equipment or capital goods that would otherwise be economically impractical. A computer, for instance, is usually far too costly to justify limited or single-purpose use. But tied into several purposely created subsystems and thus serving several different purposes, that computer—or even a more elaborate one, with additional capabilities—may find economic justification.

Conversely, the existence of a system provides opportunities for the enhancement of systems capabilities that could not apply to unsystematized operations. An example is the development of special computer centers serving great networks to allocate airline or other reservations or to control railroad freight movements. Such applications not only provide the opportunity for more intensive utilization of existing facilities, but also often help to develop the need for additional facilities.

Systems suitably equipped with and, in effect, created by linkages can provide a degree of communication and coordination not otherwise obtainable. One familiar example is the Stock Exchange ticker, which transmits information simultaneously to thousands of installations from a single typing. But when intercommunication, however routine, must involve the intermediation of humans, there is a time lag from the moment the decision is made to initiate the message until the message has been received and can be acted upon. Even with sender and receiver on an open telephone line or on a closed television circuit, there is an appreciable passage of time between the availability of the information at the transmitting end and the completion of recording or compliance at the receiving end. This is a serious matter when there is a substantial volume or continuous flow of information to be transmitted.

For many purposes this can be accomplished automatically and much more rapidly and accurately without human intervention. In such installations the effect is virtually instantaneous; often there is no appreciable delay. Such systems are said to function in real time, and they offer the possibilities of many advantages in the direct control of ongoing processes of all kinds and in the effectuating of inputs and outputs of non-physical systems or the information subsystems of physical systems.

The contribution of physical systems to economic productivity is basic and familiar, but a brief discussion is in order.

In preindustrial times the clothing worn by the members of a family was likely to be the product of that family's own work, as is still the case in some rural areas of undeveloped countries. The fleece was sheared from sheep in the spring. The wool fibers were carded, combed, spun, dyed, and knitted or woven—all without benefit of any machinery more complex than spinning wheel and handloom. The work was done in batches, and enough fleece would be processed to make cloth, say, for one garment. This quantity of wool would be advanced through the successive stages of processing whenever the lady of the house had the time for it. Some women produced more and better fabric than others, and the differences were due to individual care, efficiencies, or skills.

When the power of the steam engine came to be applied to the production of wool cloth at the beginning of the Industrial Revolution in England, it was clear that only continuous production in large quantities could justify the investment in labor-saving machinery.

The crude systems of the early woolen mills were aimed at maximizing machine productivity, at whatever—and however short-sighted—cost in human values.

The Enclosure Act, which empowered landowners to withdraw land from cropping and devote it to pasturing sheep, and other factors had brought about a disproportionate supply of unskilled labor; and it was, of course, the exploitation of these unfortunates that gave rise to Karl Marx's apocalyptic concept of class war.

Thus the discovery that a small child, with his tiny fingers, could be trained to tie knots in broken threads especially well—quickly and cheaply—led incidentally to the heartless exploitation of child labor. But it also led directly to specialization of task in an industrial operation.

The advent of the railroads and their growth and proliferating interconnections created both the opportunity and the need for the development and application of new systems techniques. Thus technology developed subsystems for signaling and communicating, provision of fuel, switching, and the like. Tasks became extremely specialized, and all operations were subordinated to functional schedules.

The availability of rail transportation in turn fostered the growth of industrial enterprises that could now exploit more accessible materials, supplies, and labor, and could reach more extensive markets. Organizations became larger, and the growing competence to manage larger organizations contributed to the development of systems. The telegraph, the telephone, and later the truck, the wireless, and the airplane made possible still other spatially dispersed service systems that in turn facilitated the ever greater development of productive systems over ever broader reaches. Thus the economy was able to grow as no economy had grown before.

The transportation systems facilitated or even made

possible the functioning of the physical systems over distances. The communication systems enabled the physical systems to be managed from remote centers. Other systems evolved, such as mail-order merchandising, which depended for their functioning on both communication and transportation systems. And with the size and growing complexity of the management of the physical systems came the recognition of need for systems approaches to the growing nonphysical operations of record keeping and services in support of management.

Because the free enterprise system rested upon the basic objective of profit to provide incentives for capital and labor and to generate new capital, economic considerations tended to enter into more decisions, policies, and initiatives. Much information was necessarily required for adequate evaluation of economic factors. This requirement constituted one of the most compelling forces behind the development and extension of information systems.

To meet the requirement, nonphysical systems were created as essential adjuncts to the operation of physical systems. The systems of paperwork proliferated into systems for processing the relevant information, dispersed in the great mass of records, to provide meaningful information in usable form to enable managers to perform their functions effectively. (The outstanding precedent for this was the indexing and reference systems used in public and university libraries.)

Advantages of Systems to Individual Participants

Advantages offered by systems to individual participants must be considered in relation to the needs and values of the participants, because in at least some situations what may be thought by some to be advantageous may be felt by others to be disadvantageous. Operational disadvantages of systems are considered in Chapter 4, and certain implications for personal involvement in systems are discussed in Chapter 6.

Physical systems can increase the productivity of labor, sometimes by great ratios. They thus provide the economic basis for increased compensation.

Physical systems usually also affect the working environment—the physical conditions of the work place. On the whole, such changes are now likely to be favorable, often enabling the worker to move farther away or be separated from unpleasant conditions, or improving the conditions.

Many systems introduce requirements for special or new skills for their operation and/or maintenance. This can provide desirable employment opportunities not previously available. A striking example of this was the apparently insatiable need for computer programmers—an occupation virtually unknown a decade or two past. On the other hand, the introduction of a system may make obsolete hard-won and previously well-paid skills.

Systems are generally most relevant to large organizations and facilitate even larger organizational size. Larger organizations tend to offer more of certain conditions than smaller organizations. Some of these conditions are considered advantageous by some workers. Such conditions include anything from membership in large, active union locals to a sense of status through employment by a well-known company, or from opportunity to participate in all sorts of employee recreational activities to superior fringe benefits, or from training and educational opportunities to the purchase of merchandise at a discount. (Of course, some smaller organization may offer some or all of these advantages, and not every large one does.) On the whole, it may be said that the indirect effects of systems on individuals via organizations are so pervasive and extensive as to be scarcely numerable.

Systems are generally relevant to complex organizations and can make feasible the evolution of even greater complexity. Organizational complexity is conducive to diversification of functions. Systems are conducive to a wider variety of specializations. Under such conditions, some employees find advantageous opportunities for their skills, experience, or potential capabilities and others, especially those of limited capabilities or aspiration, find a welcome simplification of task.

The various putative advantages suggested above, and others of which they are illustrative, are generally relevant to those in the lower echelons of the organizational structure—by far the more numerous group. However, it is of the nature of systems that they consolidate or integrate or unify a number of activities, processes, or functions. (This can also occur through the introduction of larger capacity equipment, through integration of processes, and so on.) This tends to consolidate, integrate, or unify the managerial activities related to these processes or functions. The general effect is to increase the importance of managerial effects, and thus the importance of the management function and so of the manager. (This refers, of course, to the individual actually exercising managerial functions of a high order.)

As a corollary, the installation of systems, through centralizing control, can and often does tend to move responsibility and authority higher up in the organizational hierarchy. This puts more power in the hands of the members of top management, along with the systems-provided tools for the more effective exercise of that power. The relatively few participants who find their potentials enhanced often regard this as advantageous to themselves as individuals and seek such advantages as fostering their personal careers.

Systems, therefore, incidentally or purposefully, affect the functions of the levels of management between the worker and the policy-forming levels. Members of middle and lower management often find less opportunity to affect policy and less leverage in their decisions but greater need and more urgency for administrative activity and for decisions at interpretive levels. This generally results in relative rigidity of function with prescribed patterns of performance and reporting. Some proportion of foreman, supervisors, and middle managers prefer such conditions, find them more comfortable, and so regard them as advantageous.

Of course, an extremely wide spectrum of benefits and advantages accrue to individuals through the economies and efficiencies of systems and through the products and services that they facilitate or make possible or more economical. The vast reach of these basic benefits—derived by most of us from pervasive systems, merely by living in civilized communities and enjoying a fairly widespread standard of living—has a beneficial fallout affecting the majority of the population of industrially advanced nations.

Many systems provide what amounts to extensions of the personality, especially those systems providing communication effectiveness, such as the mail, the telephone, and the like. Others, such as plane, rail, and bus systems, provide great personal mobility, as do the systems that make the automobile practical, such as highway systems, fuel distribution systems, and parts and repair facilities.

Individuals benefit from systems that produce and distribute and from those that serve. They benefit from participation in organizations that depend on systems. It is not too much to say that industrial civilization, with all its benefits and drawbacks, is derived from and dependent on the development of systems that make possible the generally advancing level of exploitation of scientific knowledge and of natural and human resources.

The necessity we all face in the Age of Systems is not only to exploit and extend the advantages of systems, but also to become more aware of their disadvantages and undesirable effects and to take action to minimize them.

4 Disadvantages of Systems

A major and primary disadvantage of systems is the all or nothing characteristic that they impart to much organizational functioning. Some degree of dependence of organizations upon the internal components or external resources that enable them to function is inevitable, but when this amounts to or verges upon total dependence, the organization is indeed giving hostages to fate. When functions are assigned to individuals and individuals are sufficiently interchangeable, the organization carries a traditional safeguard—so long as replacements are available. But when it places its dependence upon a system and the system breaks down, the organization may face a serious if not fatal emergency. Similarly, if a system is dependent upon a set of required conditions (which may or may not consist of or include a system or systems), that system is vulnerable.

Functional Dependence

For the purpose of discussion, we may classify the relationship of systems (or other potentiating conditions) to the organization as totally external, external to the activity concerned, or internally functional. Organizational functioning thus becomes dependent upon conditions required by an essential system.

Systems totally external to the organization and upon which it may be extremely or totally dependent include governmental, economic, social, public utility services, transportation, communication media, availability of essential supplies, and the like.

As examples, organizations have been drastically affected by public disorder, curfews, and martial law, resulting from breakdowns in the system for maintaining law and order. Abnormal conditions in the financial systems, such as the bank closings in 1933 and the credit restrictions and extremely high interest rates of 1969–1971 have caused grave difficulties to many organizations. Interferences with the operation of such systems as postal services, railroads, trucking, airlines, and similar essential services by strikes, natural catastrophes, accidents, or otherwise have repeatedly interfered with regular organizational functioning. Stoppage or deterioration of telephone service can have a highly obstructive effect. Power failures can bring total stoppage of manufacturing operations, or practical inaccessibility of offices on higher floors of tall buildings, or other unsurmountable difficulties. Failures of the systems supplying fuel or required materials can halt physical systems completely. There are any number of other examples of the effective dependence of systems upon the required functioning of external systems or of adequate external services or conditions.

Meeting and overcoming the difficulties and obstacles posed by such conditions usually involve recourse to alternative resources, the need for which may or may not have been forehandedly anticipated. In some instances, of course, alternative systems may be involved. But the larger, more comprehensive, and specific the system, the greater is its probable dependence upon a standardized set of conditions, and the less practicable it is likely to be to provide alternatives that will allow effective functioning. Nonsystematic improvisations will usually prove to be poor and temporary expedients.

Examples of systems external to the activity (but within the organization) are an air-conditioning or a heating system that affects the physical environment to a point of dependency or a system handling undesired by-products, such as smoke or fume treatment, effluent disposal, or pollution control. Such systems are not essential to the functioning of the primary system, but without them—for health, legal, or other reasons—the primary system may not be operable. The availability of programmers or card-punch operators may affect the operation of clerical systems. A strike of refuse-removal workers may cause such a backup of used paper in a large computerized clerical activity as to interfere seriously with operations. On a smaller scale, a failure in the maintenance system to provide a cleaning service for a battery of special printers resulted in nearly illegible printouts on punch cards, leading to an increase in errors and decreased productivity among clerks who were required, in several systematized processes, to read them swiftly and accurately.

Breakdowns for any reason within an integrated functional system usually lead to complete stoppage of the system. A shortage in the supply of a single part can bring an assembly line to a halt. The absence of a mechanic can stop a machine, which can bring a whole series of processes to a halt. The failure of an instrument or a valve can affect an extensive process in a refinery or a chemical plant. Of course, excellence of design tends to eliminate avoidable bottlenecks, and redundancy—backup, emergency and spare equipment, personnel, or available capacity—can take up the slack when adequately provided. But when such backup facilities are idle, they constitute an inefficiency in the system, justified only as insurance or to facilitate maintenance.

Similarly, the breakdown of a computer can bring a number of clerical or other systems to a halt. A single error in a computer instruction can throw a paperwork system into chaos. The sickness of one critically placed specialist can hold up the work of many.

The pervasiveness of systems and their highly involved interrelationships, concatenations, and interdependence greatly magnify their vulnerability. Some systems are so extensive that their malfunction or dysfunction have extremely wide-ranging effects. Examples would be the postal strike of 1970 and the power failure in the Northeast in 1965. Some systems have peculiarly strategic relations to others. The slowdown of plane controllers in 1970, though relatively small numbers of people were directly involved, affected traffic in and out of major airports and had wide-ranging effects on the operations of many airlines. This in turn affected the movements of many thousands of individuals and of untold quantities of mail and goods, and these consequences in turn had their myriads of repercussions, large and small, thus again multiplying the effects exponentially.

This increase and expansion of vulnerability is the price that is often paid for the advantages of systems. The obtainable advantages are undoubted, and it is logical to seek their maximization. At the same time, the possible and—in a general, probabilistic sense—the inevitable consequences should be recognized, and a reasonable provision should be made for anticipating them and minimizing their adverse effects.

When elaborate systems are designed and installed, and less dependence is placed upon humans and more on machines, there is a natural tendency to deprecate the likelihood of failures and breakdowns. Too often, this results in inadequate, unrealistic provision for eventualities. Furthermore, failures in a system generally call for measures not provided by the system itself; they often demand improvisation, initiative, and mechanical or human resources not required by the system itself. To the extent that the existence and the functioning of a system militate against the availability of the means required in the event of inadequacies, malfunctioning, or stoppage, a liability exists and may develop further. If such means are indeed unavailable when a need arises, many of the benefits of the system may be wiped out, in the balance.

Classic examples of the limitations of systems and of the need for unsystematically versatile backup and supplementation are seen at most technological forefronts. Typical and outstanding is the series of Apollo moon shots, where the most advanced of physical systems are controlled by the most elaborate and complex of nonphysical systems, but with hundreds of human components to provide the evaluations, linkages, and other operational essentials that cannot be built into the system.

Every now and then an airline pilot, facing some emergency (often a breakdown in a system), is called upon to exert personal resources of skill and knowledge, the exercise of which, however well provided for as a systems backup for potentially catastrophic contingencies, is certainly not a matter of operational routine.

Human linkages in communication systems—from telephone switchboard operators to interpreters, from receptionists to representatives, from secretaries to colleagues—contribute all sorts of unsystematic inputs to communication processes, some constructive, some otherwise. The importance of the individual contribution can become decisive, either way.

Marketing and selling activities usually involve many nonsystematic operations, though direct mail operations, for instance, can utilize highly systematic selling activities by eliminating the face-to-face element. Most selling involves face-to-face interactive processes that can be only partially systematized, because salesmen cannot be expected to behave in a completely uniform manner nor can their customers' behavior be standardized. However, some marketing activities approach systematization. Prime examples are to be found among door-to-door vending operations, which attempt to standardize representatives' performance, and self-service merchandising activities, which attempt to minimize the human component. These offer interesting examples of human subordination to systems but require careful and skillful management.

Increasingly, it may be generalized, marketing tends to operate on a probabilistic basis. Mass distribution of standardized products, calculated on the basis of meticulous market surveys and fostered and supported by saturation advertising campaigns, is based on statistically determined expectations of potential results. Comprehensive plans are made, utilizing sets of systems for their implementation, to effectuate the established goals.

When deficiencies, shortfalls, or other disappointing deviations from plan are encountered, remedial or supportive action is usually taken to counter the adverse developments and to provide reinforcement where the need is indicated. Thus various elements of the overall system, including component or contributory systems, are manipulated within the available resources. When favorable results are obtained, manipulation of major components takes place to emphasize and second the effective elements of the program.

Ultimately, a relatively stable situation is hoped for, usually with planned increments, and marketing management operates to maintain at least the projected level of effectiveness through the use of market information and the adaptive utilization of its resources. Such situations lend themselves to and even require the exploitation of system operations. Obviously, all such systematized operations are extremely if not totally dependent upon the reliability of the survey data and the inferences drawn, and tend to fail if the data on which they rest prove to be erroneous assumptions rather than fact.

Probabilistic techniques, leading to system applications, are used in many other organizational activities, often through computer technology. Plans based upon the results are, of course, only as sound as the assumptions upon which they rest, and here time can introduce changes—sometimes quite suddenly—that invalidate the most carefully laid plans and make irrelevant the best designed of systems.

Form Versus Content; Rigidity

A distinction must be made between systems that provide a general service and those that perform only a specific function. Thus the telephone system provides a service that is general in that it can and does transmit any conversation on any subject between any two telephones in the system. A transportation system can and does carry virtually any passengers and/or freight between any two locations within the system. A power grid with its distribution systems can and does supply electricity at any connection, regardless of purpose, save for technical considerations.

The telephone system disregards the messages of those who use it, offering only the specific facility for and service of transmission. The carriers likewise do not propose to change the persons or things they carry, except by transporting them. The power company ignores the uses made of the current it provides, except as this may affect demand.

To the telephone company, message content is of no importance as input or output or, rather, all inputs and all outputs tend to have similar value, transmission of all messages being charged for on the same gross basis. Idle gossip, flirtation, or unwelcome sales solicitations have equal claim with calls critical to the consummation of major transactions or urgent personal affairs. (The state of New York only recently passed legislation providing priority on party lines for such emergency matters as reporting fires or calling doctors.) To the carrier also the individual passenger's reasons for traveling are immaterial, as is the passenger's personality; and the contents and value of transported goods are irrelevant, except for purposes of rate setting and liability. The electric power systems are concerned only with what can be delivered over the wires. Their inputs are the energy of heat or of falling water; their output is in transmitted wattage.

The telephone company measures its services in the number, duration, and distance of calls. The carrier measures its effectiveness in passenger miles and ton miles. The electric utility measures its activity in kilowatts delivered, and one kilowatt is, as nearly as possible, exactly like every other kilowatt.

However, a majority of systems require a narrower range of input, and actually operate upon the content, affecting change. Thus manufacturing systems require inputs of specific components or materials, and nonphysical systems usually require inputs of data in processible form.

With manufacturing systems, the specificity of output increases with successive stages. A metallurgical plant is designed for the input of a specific class of ore; the output, a refined metal, can be put to a variety of uses. But as the metal is treated and fabricated within a system that produces assemblies—let us say automobiles—as output, its end use becomes limited to incorporation, as planned, into an assembly. Thus a system in Chile converts the copper content of huge quantities of ore into anode bars. Electrolytically refined, the copper could serve a myriad of ends. But once it becomes an input into a plant manufacturing automobile radiators, it can serve no purpose other than that specifically intended: The system will continue to process it into a component of the cooling system of a particular make and model of car. It is useless for any other purpose, and if it cannot be used for that purpose, it becomes scrap, to be remelted and refined in order to be restored to the basic condition in which its potential uses are open-ended.

This narrowing and focusing of output applicability is typical of the advanced stages of most physical systems and is inherently justified as serving the purpose underlying the system's existence: to produce a usable product or class of products or a product line. Some of the effects on the individuals involved are considered in Chapter 6.

These observations about physical systems are, to a considerable degree, appropriate also to nonphysical systems. A great majority of these embody primarily clerical operations that process information in order to produce information outputs for specific purposes.

In typical examples, a variety of relevant personnel data—names, Social Security numbers, hourly wage rates, hours worked, authorized and required deductions—are fed into a system that produces payroll data or even writes the paychecks. Names, addresses, charge account numbers, items purchased, charges, and taxes are converted into monthly bills. Records of withdrawals, checks honored, deposits, charges, deductions, and credits are assembled into monthly bank statements.

In all these examples, simple factual records and specified factors are processed into a form intended for a specific purpose. The process in many systems operates automatically, or virtually so, from input to output. It is designed and programmed to produce a specified output from specified input. By definition, it cannot cope with exceptions.

This kind of rigidity results inevitably from the sacrifice of generality of applicability for maximum effectiveness for a narrowly defined purpose. An individual can be trained or a tool or a system can be designed to meet or fulfill a single pinpointed need or function more effectively and more economically than to meet a number of needs or fulfill a broad spectrum of purposes. Technology facilitates specialization; economics fosters specialization. Therefore, many system applications call for highly specialized systems. And such systems, as they handle suitable inputs with more and more efficiency and economy, tend to have decreasing capabilities for handling exceptions, and any tolerance tends to become increasingly marginal.

As an example, a firm receives one thousand items of incoming mail daily. These can be sorted for distribution in a process of one phase or a very few phases—without opening the envelopes. But the processing from then on must depend upon the nature—the content and purpose—of each item. An order will be processed as such; a request for payment must be processed differently. Now let us suppose that nine hundred of the incoming items are orders. A system is devised, based upon a standard order form, to carry the input. Orders received on the standard order form enter the system directly, but orders received otherwise, as by letter, must be transferred to a standard form before they can enter the system. A clerk is detailed to this task, and the system functions.

As time goes on and the system becomes thoroughly established, more and more orders appear on the required form, fewer and fewer appear otherwise. At some point the work load of the clerk who is detailed to transcribe orders onto the form becomes extremely light, and the clerk is given other regular work. These new duties soon become paramount, and thus the transcription of orders becomes the exception—a distraction from regular duties.

These exceptions then are set aside, perhaps, until enough accumulate to alert a supervisor or even until the ensuing delay arouses a customer. The system is not geared to the exceptional input.

An outstanding example is that of the bank check. The introduction of magnetic ink imprints on checks made possible an extensive system that can process the checks from deposit, through clearing, to charging against the account drawn upon. In the beginning, this system was used along with untreated checks, which were processed by existing methods. But as more and more checks were imprinted with the magnetic code, an ever smaller proportion appeared without the coding that made the system possible. Soon the untreated checks began to be considered exceptional, then as nuisances. Now they are often declared unacceptable. The system has gained efficiency, but it has become more rigid: It has narrowed the specifications for its input.

Although the logic of such developing refinement is often clear, some of the results are more serious than are generally realized. Increasingly, employees serving a system are not available to service exceptions in the input, and if necessarily made available, they are all too often ignorant of the necessary procedures. As time goes on, the older employees who remember the presystem operations are promoted or retire or leave, and there is little or no resource to back up the system.

When the system itself involves human inputs, as with the processing of items in paper flow lines, some error is inevitable. But the system itself seldom provides the means for correcting such errors or overcoming their effects. This must usually be accomplished supplementarily by competent individuals assigned to or specially called upon for such tasks. But the necessary knowledge, broader understanding, and versatile skills required to provide remedial services are seldom if ever developed by performing a specialized function within a rigid system.

Another aspect of this kind of developing situation is related to the extraneously imposed rigidities (jurisdictions) of some labor unions, which define what may and may not be done by members of each union or each employee category within the union. Related to this is the division between craft unions, with a generally broader range of recognized functions, and industrial unions, which often stress narrowly defined job descriptions. Regardless of any justifications offered, it seems beyond doubt that all such jurisdictional and definitional interjections have the effect of introducing more rigidities into the situation and thus further constrain the operation of systems.

Thus rigidity, specialization of function, and narrowed acceptability of input are characteristics of many systems. They affect the requirements for personnel functions, and tend to leave the organization increasingly vulnerable to mishandling of (or failure to handle) more or less exceptional items of organizational intake.

The inevitable result is familiar, for instance, to many thousands of retail customers who find it all but impossible to have errors in their charge accounts corrected. Patrons of brokerage firms or mail order houses, depositors in banks, users of credit cards, subscribers to magazines, and many other categories of active participants in the economy have found themselves involuntarily involved with stubborn discrepancies in their accounts that the rigidities of the systems seemed determined to perpetuate. Examples are too numerous in the experience of so many to require further illustration of the extent of the consequences of system rigidity.

Downgrading of Participants

Up to the late nineteenth century, footgear was made entirely by hand, and the bootmaker or cobbler was a skilled artisan who could take the measure of the feet to be fitted, cut the leathers, form and shape the shoes, and then peg and sew them into finished articles. If there was division of labor or specialization in his shop, it generally took the form of letting an apprentice do most of the routine sewing or pegging, once the positioning of the stitches or pegs was clearly indicated.

The bootmaker was a man respected for his skills and for his individual contribution to the local economy. His personal competence was a matter of discussion, sometimes of pride, among his customers. He took his place among tanners, fullers, silversmiths, millers, blacksmiths, carpenters, masons, and other craftsmen essential to the community.

When the mass production of shoes was initiated in Rhode Island, the entrepreneurs mechanized the process of making shoes by establishing standard sizes and separating the manufacturing process into steps simple enough to be performed on a machine. They provided almost one hundred different machines, each one designed to mechanize one step in the total process of converting leather into shoes of standard sizes. They could hire willing workers who had no experience—much less skills—in making shoes, and they could readily train each one to operate one machine, to perform one operation in the overall process.

They had a system, and the system replaced the skilled bootmaker with the unskilled machine tender. The cobbler whose business disappeared in the ensuing competition could seek a job in the shoe factory, but not as a cobbler—only as the servant of a machine.

Every technological gain has brought changes affecting those whose lives were in some way involved in the displaced status quo. When the technological change makes obsolete the occupation formerly provided to an individual under the status quo, the individual must necessarily seek another occupation. If the new occupation requires little or none of the skills and competence the individual used in his former occupation, there is a downgrading effect—unless other or new requirements for comparable or greater skill and competence exist and the individual is enabled to meet them. This is seldom the case when a craft is replaced by a system, and in this sense systems tend to have a downgrading effect upon most individuals involved.

The economic consequences of the transition may be beneficial to the individual (he may earn more as an employee working within a system than he did before); and the technological consequences are usually considered beneficial in a material way to the economy generally, as well as making products of new technology available to the individual; but the occupational consequences are quite another matter for a majority of those directly affected.

The psychological and sociological effects of functioning within a system are discussed in detail in Chapter 6.

However, it is obvious that no one can perform one standard, relatively simple task as one stage of many in an overall process, remaining relatively anonymous and readily replaceable in that task and one of many on a low level in a large organization, without having it affect his own concept of himself and his place in the world, and his subjective evaluation of the social, cultural, and economic environment in which he does his daily work to earn his living.

The machine has taken over many disagreeable, difficult, and even dangerous jobs. It has made possible many operations that would otherwise be impossible, and it has made others economically feasible for the first time. It has made vast contributions to the ease and material enrichment of the lives of many. And by liberating man from much labor and by giving him otherwise unattainable freedom of movement, it has enabled man to enrich his life in other ways. Furthermore, the ever-extending potentials of the machine are a constant challenge to those who work toward realizing these possibilities and their future applications. And systems have greatly extended the utility of machines.

But there is no action without reaction—so runs an immutable law of physical nature. Many would extend this concept to the intangible world of spiritual and moral values and to the emotional world of human needs and satisfactions. Certainly, with many machines and systems there have been great gains, but the losses are also apparent to some—though often obscured by an acceptance of the inevitability of progress.

The kind of social and psychological effects discussed in relation to the machine are multiplied when individuals become components of systems.

The questions that remain are those about the inevitability of the adverse effect of progress. Is it unavoidable that a technological advance will result in a loss of human values—moral, spiritual, or emotional—to balance or partially offset the gains in human benefit and satisfaction? Is it inevitable that more and more individuals will find themselves actively involved within systems? Is it inevitable that systems exert the adverse environmental effects that have so often accompanied the transition to human participation in systems? And if these developments are not inevitable, what can be done?

The physical system generally multiplies the individual effectiveness of most of those who are employed to make it operate, by applying their efforts to more productive equipment. In many operations, this applies to a minority of workers only, often a small minority. The moment-to-moment functions of many employees are left almost unchanged, or even downgraded to near automatism.

When highly mechanized equipment is introduced into clerical operations, its contribution to increased productivity may be accompanied by a variety of effects upon individuals. A copier relieves a typist of tedious copying. A better typewriter, adding machine, dictating machine, bookkeeping machine, or calculator can help greatly to improve individual productivity. But some equipment does so at the sacrifice of personal skills or elements of skill such as typing or shorthand, or of some forms of personal contact such as dictation, or of variety of work when the individual becomes identified with and attached to a particular machine for specialized purposes and tasks.

As physical systems grow, the maintenance and repair of the equipment that embodies the system is unlikely to become more exacting for on-the-job employees; it is more likely to change toward the far less exacting process of simple replacement of interchangeable units, not merely parts, but complete units such as subassemblies that are the products of systems in the suppliers' plants.

Most of the individuals who serve systems increase their productivity only in the sense that they are contributing to more productive systems, which do not generally require very much, if anything, in the way of increased quantity or quality of individual contribution. The pilot of a superjet may or may not need somewhat greater knowledge and skill than the pilot of a smaller plane, but it is doubtful if many of the hundreds who support him will be required to have significantly higher capability.

A parallel situation appears to exist with most organizational or information systems. The executives who receive the data and draw upon it for planning, solving problems, and arriving at decisions may be vastly advantaged, but most of the clerical employees down the line are little conscious of such ultimate advantages deriving from their increasingly dull routine.

Thus we see a functional polarization taking place. Those who can utilize systems effectively are enabled to increase their managerial potency; they become more significant factors in their organizations. At the same time, many or most of those who make this possible by participating in the functioning of the systems themselves find they are engaged in increasingly stultifying tasks, without inherent interest, yielding no satisfying sense of real accomplishment, and too often leading nowhere.

The division of labor in a physical system involves one or more series of operations in the course of processing, manufacturing and/or assembling a salable product. While any one participant's assigned task may be tedious, unexacting, and unsatisfying, at least the worker can usually identify the end product and understand its use and value. But the component processes of information systems are often fractional, highly abstract, and of extremely transitory character; the physical entities dealt with are of no inherent value and rapidly become wastepaper; and the end uses of the system may be esoteric, incomprehensible, or understood in so vague or general a sense as to be without interest. Thus the ultimate utilization of the individual's efforts have neither meaning nor meaningfulness to him.

Multiplication of Failure and Error

Compared with the functional units that they replace, systems are generally much larger and are often unitary in effect. Too often these developments compound a failure and multiply an error. They may or may not increase the probability of adversity, but they do increase the magnitude of the potential loss.

In physical systems, quality control practices are often employed to eliminate or to minimize the occurrence or effects of adverse characteristics in the materials and products processed. In general, these are aimed at early detection of deviations from standard so that deviants may be rejected from the system or corrective measures taken.

In physical systems, it is generally recognized that a failure in one phase can result in total stoppage. The usual measure taken to meet such risks, besides preventive maintenance, is the provision of substitute, alternative, or backup facilities—sometimes whole systems in themselves. In some cases, these operate automatically, or can be quickly brought into action. But in other situations, much valuable time is lost while a substitution or repair is made.

In physical systems generally, directly related economic factors determine policies and managerial decisions. The long-term profitability of a system must outweigh the costs of its occasional failures. Insurance can be carried to cover some risks. Some work can sometimes be shifted from one line or shop or plant to another. Some systems have a capability for some degree of flexibility or conversion, especially when components are multipurpose (some machine tools) or flexible (some employees).

Nevertheless, the increasing size and scope of many systems and systems-involved facilities are enlarging some risks and increasing adverse possibilities to a degree that surely approaches—if it has not reached—the unacceptable.

As an example, consider the modern tanker ship—an apparently indispensable feature of our dependence upon intercontinental commerce in petroleum. The tanker is an integrated cluster of systems—propulsion, navigation, steering, storage, pumping, internal transfer, and others. In routine operation, it makes little demands on any member of the crew for nonroutine performance. And the economics of tanker operations have for some time all but guaranteed profitability.

The increasing demand for tankers resulted in new technology, which greatly increased size and provided greater automation of onboard operations. But the increased profitability of the larger vessels is counterbalanced by at least two adverse factors. Delays of the same length are far more costly with the newer larger tankers than with the older smaller ones, but they are at least as likely to occur and are perhaps more difficult to overcome if they involve repairs. In addition, the increased capacity increases the extent of possible damage and risk in the event of a loss of cargo at sea, and the world's increasing sensitivity to seacoast pollution (since the *Torrey Canyon* poured oil on the beaches of Cornwall and Brittany) greatly enhances the effective financial liability.

At this writing 500,000-ton tankers are on the way. They will carry more than twenty times the quantity of oil carried by the tankers of the 1940s, and probably with little increase in the size of the crew. They will operate within many present systems for oil transportation, and for loading and unloading at offshore terminals. They will be able to provide great operating economies in normal operation. But to do this they must demonstrate an overall dependability far greater than that of the average smaller tanker. One accident or labor dispute or tempest or navigational error spread among ten or twenty or more units would have incapacitated only 10 percent or 5 percent or even a smaller percentage of the total. And a half million tons of oil spread upon the sea could set a world's record for catastrophic pollution. Thus considerations quite extraneous to the advantages of a system may militate against it, and sometimes the limitations must be imposed from outside the organization.

The efficiencies and economies and the many advantages of systems make it appear logical to extend them to the maximum feasible degree. But the arguments of technological and economic feasibility and of material advantage are often opposed by considerations of an entirely different kind. Only two illustrations will be offered here, but there are many others.

In the United States many systems exist for listing and maintaining data on individuals. Credit information is extremely pervasive, being available from banks and credit bureaus and by mutual interchange of credit information among companies. Individuals are listed according to all sorts of indexes, and anyone can purchase lists of individuals by specified category. The mails are filled with advertising addressed from such lists. The states maintain (and some of them sell the use of) lists of automobile owners, drivers, and holders of other licenses. Local governments list owners of taxable property. Many companies maintain lists of their customers, the use of which they offer to others for a fee.

In addition, the local police record the names of those arrested; the local and state law enforcement agencies maintain extensive lists of those with records and others; and federal agencies, notably the Federal Bureau of Investigation, maintain enormous and elaborate lists. The Social Security records cover every individual who earns or has earned money. The Internal Revenue Service has a great deal of information about everyone who pays an income tax.

There are proposals to extend these coverages to a logical extreme: Let everyone be listed from birth in a comprehensive computerized file maintained by the federal government. Universal fingerprinting is often proposed as a logical concomitant of this plan; the fingerprints of millions of Americans are already on file through service in the armed forces, or conditions of employment requiring it, or other noncriminal activities.

The advantages and benefits of such a comprehensive system are undoubted. But the objections made have so far prevented adoption of the scheme. And these objections are not concerned with technological or economic feasibility, and they discount the very real utility attainable. These objections arise from a desire to defend human values that are hard to define, but that are very real to their proponents. They speak of privacy, and of the encroachments on human liberty, and of interference with private affairs. They fear improper usage of the files. They point to the danger of their abuse by an increasingly centralized government seeking extensions of its control over individuals, or even seeking to counter or stifle constitutionally guaranteed freedom of dissent. They describe the plan as totalitarian in concept, and compare it with the practices of police states.

Obviously, many thoughtful people believe that some systems can be carried too far.

Another kind of example is provided by the typical case of a retailer of custom clothing for ladies, with a distinguished salon in a fashionable neighborhood. When he had only a few hundred charge accounts, the bookkeeping was vertical, that is, each bookkeeper had complete responsibility for certain accounts and kept liaison with the sales clerks who invariably served those accounts. Thus the bookkeeper would know all about each of the accounts in her charge—what the customer ordered, when, and how much; what materials were ordered, when, and how much; what work had to be done, when, and the cost in wages; and when delivery was made. The bookkeeper sent out the statements and handled the receipts, so she knew if and when the account was paid and the indicated credit status of the customer.

The bookkeepers actually came in direct or telephone contact with the principals of these accounts from time to time as they handled inquiries, complaints, corrections, collections, and so forth. They felt that they knew the lady customers; they often asked to see the clothes that each customer bought; they came to take an interest in each account and so in their own work. They read the society columns for news of their customers' social triumphs, which generally included rapturous descriptions of their latest gowns.

As the business grew, the clerical load greatly increased. The owner added a line of less expensive ready-made garments. New clerks were hired and then an office manager. Eventually the bookkeeping department had to be reorganized as a new system was obviously required. The new system was largely horizontal, that is, each order or purchase was processed in stages and each clerk was now involved with all the customers, but only in one stage.

The new system was undoubtedly more efficient, and it made use of up-to-date equipment that provided advantages. But now each clerk was isolated with specific, limited functions. Gone was the sense of vicarious participation in the triumph of Mrs. Prominent's gorgeous green lamé at the Prosperity Ball. Gone was the sense of involvement in the excitement and glamour of high fashion. Gone was any basis for interest in their work, which was now a relatively mechanical processing of items in a paper flow, apparently disconnected from reality and life.

The bookkeepers who had known a different way felt definitely downgraded. They were extremely unhappy. Those who did not have too great a vested interest in their positions began to look elsewhere. The others tried to adjust to the new situation, and their adaptation involved a considerable diminution of personal identification with their work. Customers who found errors in their accounts, and especially those who telephoned about it, soon noticed a difference. And the now disgruntled old-timers contributed rather negatively to the indoctrination of the new clerks by talking rather bitterly of "the way it used to be."

The new system had an intangible cost that is not clearly apparent on the earnings statement but still is undoubtedly reflected in operational effectiveness and profitability.

Limitations of Human Involvement

There are, of course, special situations in which participants in systems become highly involved and identify strongly with the system, with its success, or with its characteristic manifestations. This is discussed more fully in Chapter 6. Matters of morale, esprit de corps, and leadership can be highly motivating within systems, though they derive from individuals and not from the system itself. Identification with the leadership or with group goals can be highly motivating in or out of systems.

And, of course, there are those who respond positively to the material rewards or status that may be associated with effective participation in systems. And there are always those who feel most at home, most relaxed, and most secure when their activities are clearly limited and prescribed and no initiative is expected of them.

But, on the whole, activity as a human component of a system almost always offers less in the way of nonmaterial satisfactions and a lesser degree of interesting experience than does involvement in less systematic situations. There are those who care about this, and some care very much. In fact, the many satisfactions absorbed in less systematic situations are frequently missed when the advent of systematic involvement eliminates them.

One kind of example of this is provided, in an extreme form, by the spectacle of a tall modern building being erected in India. An imposing edifice of perhaps twenty stories, modern in every respect, is built with extremely crude methods. The tall steel frame is faced with a bamboo scaffolding, tied together with bits of rope. Hundreds of coolies stand on this and pass along an unending series of wooden bowls, each holding a few scoops of concrete, from the street to the highest level. This primitive system works, of course, and eventually the building is completed.

The architects and engineers responsible for the design and construction are entirely familiar with available technology. The building itself will have all modern conveniences—high-speed, automatically controlled elevators, total air conditioning, up-to-date plumbing and electrical systems. Once the tenants move in, the building will not differ greatly from similar structures in New York or London. But in the course of construction, large numbers of human beings are involved in a primitive system—using human components to perform the simplest of repetitive operations—long ago replaced elsewhere by more efficient systems utilizing equipment for similar purposes.

Why are the applicable systems of materials handling not utilized? The answers are to be found only in the Indian economy and the sociology of the Indian coolie—the overpopulation of India, the misery and degradation of many millions of its people, its unemployment and hunger and pitifully low wages, and even the weakness and smallness that result from undernourishment.

By comparison, most manual workers of advanced economies could be induced to participate in such an activity only on a temporary basis for special reasons, for instance, in a bucket brigade to quench a fire.

For the pitifully poor coolie, who exists on the building site with his wife and children in a squalid encampment with no recognizable amenities, the daily task of passing the wooden bowls does not represent downgrading, but upgrading—from unemployment to employment, from starvation to subsistence, from doom to survival. If the entrepreneurs chose or were allowed to substitute modern machinery for the hundreds of coolies, this would really constitute an intolerable inhumanity. Many of these pitiful people might starve.

Considerations of systems seldom involve such basic survival issues or values. But, on the other hand, purely technological and economic values can never exist alone, and it is likely that other values—of considerable relevance—are quite often overlooked when decisions are made about the introduction of systems.

When systems involve humans primarily, their planners almost invariably rely on rigidly prescribed formulas to define the range and specifics of the accomplishment expected in the performance of each task. There are standards set for each operation, and performance must be up to standard. These usually specify narrow tolerances for deviation from the specified standards. Greater deviation from such prescription than the set tolerance is not permitted (theoretically).

However, it is the rejection of superior or extra performance that many individuals find most oppressive, stultifying, and intolerable. In many situations, individual performance must not exceed or be superior to the standards set, or the results may interfere with the planned operation of the system. In many systems, performance superior to that specified may result in the change of specification to require of all the level of performance demonstrated by one or a few.

The setting of standards for performance on the job has a major bearing on levels of compensation.

When systems involve equipment, the rigidity can be great, because the job description of the machine is designed and built in. Human involvement is usually provided for only to maximize the productivity of the machine in the mode required or to deal with anomalies or emergencies against which the machine has no automatically adaptive reaction.

Some machines are extremely flexible in use and actually serve as extensions of human capability, so that individual energy, skill, or initiative in their use can make a significant difference. These can range from bulldozers to microtomes, from jackhammers to airplanes, from milking machines to calculators. But as the utilization of such machines becomes more effectually absorbed within the functioning of systems, the occasions and opportunities for initiative tend to disappear and requirements for skill tend to be limited to ever narrower channels.

Thus systems demand the presence of individuals and the performance of well-defined functions. But many systems increasingly reject all capabilities and interests of the individual superfluous to the performance of the required tasks. They impose ever greater limitations, both on the number of humans involved and on the degree of personal involvement for each individual.

Systems and Intangible Values

The basic concept or idea of "system" that underlies the way many Americans think about themselves, their country, and the world has developed in parallel with history, and especially with the economic and social development of the United States.

The late years of the eighteenth century and all of the nineteenth century were characterized by the individualism of those who were able to take the drastic step of leaving their ancestral homes and migrating to America. Many were driven by aspirations to build fortunes impossible to do in Europe; and they thought in terms of hewing out their own homesteads, setting themselves up as independent tradesmen, opening their own shops, becoming merchants, ship owners, planters. They thought of themselves as masters of themselves, determining their own lives, being in control.

But the growth of organizations made it difficult for the individual to survive in many lines of endeavor. Society became more stiffly structured; the development of technology made specialization, and therefore the integration of diverse capabilities, inevitable; and the rising tide of systems made it all the more difficult for the individual to manifest more capability than the system requires.

The crescendoing effects of governmental measures and the progression of federal, state, and local controls added to the overwhelming realization that the individual is largely helpless within the great, all-encompassing socio-economic, cultural-legal system. Relatively few people try to create their own "thing." The relatively large number who fail are usually reminded that "you can't beat the system."

So the individual becomes an attendant at a service

station of a vast organization instead of working on a farm. He enters a factory instead of learning a trade. He becomes a clerk instead of preparing himself to command a ship. He graduates from college to become a trainee for a subordinate post. He becomes a salesman instead of a merchant. He thinks of himself as an employee, not as an enterpriser. He accepts the system.

The more he succeeds within it, the more he is truly a part of it and helps to perpetuate and expand and intensify it. As an individual, he is cumulatively dominated by the forces and influences he knowingly and unknowingly accepts.

Political leadership may actually be a dynamic moral force exerted by a dedicated, utterly convinced individual, acting upon the reason or the long-term aspirations of those who follow.

Or political leadership may be a dynamic emotional force exerted by a passionately sincere individual, acting upon the needs and hopes, but also upon the resentments and the suspicions, the prejudices and the bigotry, the fears and the hatreds of followers.

Or political leadership may be of the nominal kind, derived from the reflexes of the political hack who is ever sensitive to the tendencies of his public so that he may "lead" them in the direction in which he believes they want to go.

We have, also, of course, another class of aspirants to political power, who may use all the skills, sensitivities, and craft of the hack, and perhaps some of the demagogy of the bigot, but who, once he attains power, tries to go his own way.

In the Age of Systems, many of the factors affecting and serving political leadership have so altered the dynamics as to change drastically the nature of politics and even the direction and control of government. Because the processes of politics and most of those of government function through communication, we may consider the impact of systems upon the two essential parallel levels—the mechanical and the substantive. The first level is concerned with communication processes, the second with communication content and its evaluation.

On the mechanical level we need only consider that the founding fathers produced the Declaration of Independence and the Constitution by writing with quill pens and blotting ink with sand, and that their campaigns for the most urgent issues were largely a matter of writing letters, publishing the literate essays of *The Federalist*, and talking with small groups of opinion leaders in taverns and coffeehouses. While letterpress printing was available, there was no typewriter, no mimeograph, no instant offset, no photocopiers, no sound recording, no photography. There were also no planes, railroads, or cars. If Sam Adams in New England wanted to address a crowd in the South, it would cost him weeks of travel by horseback, coach, or sailing vessel. The telegraph and telephone, not to mention radio and television, had yet to be thought of. There was not even an effective postal service. So there was no mechanical basis for communication systems and little occasion for more than skeletal organization. And there were no true political parties until decades later.

The personalities of Washington, Jefferson, Madison, Franklin—even of Lincoln, almost a century later—were all but unknown to the general public. The means for expanding and extending the personality were limited to the resonant sounding board of a church or public hall and to the range of vision of a bystander in a crowd.

The constitutional debates were conducted in Philadelphia without an amplifying system. And because the Gettysburg Address was not carried by a public address system, how many persons could possibly have heard it as Lincoln gave it?

Orators had to be strong, healthy, and possessed of powerful voices if they were to be heard directly by more than the nearest in a crowd, because voice amplifying systems were not a significant factor until the 1930s. The radio was only the toy of a few in the 1920s, and unimportant politically until the day of Franklin D. Roosevelt and his fireside chats.

Television began to develop its public in the period following World War II, but Adlai Stevenson, with all his undeniable charm, was unable to convert his broadcasts into sufficient votes to prevail against Dwight D. Eisenhower. In fact, TV was not generally regarded as a truly potent political tool until its effective exploitation by John F. Kennedy. It is probably quite significant that the first president to undergo almost universal exposure on the newly omnipresent tube, Lyndon B. Johnson, so forfeited public confidence that he felt forced to withdraw from the arena.

The coming of the Age of Systems has introduced vast changes into the political scene. Some of these are so apparent that they tend to obscure the others. The obvious changes are on the mechanical level. We see the mass media, and especially TV, dominating the campaigns, with unavoidable consequences for the importance of money to buy network time. A larger TV budget may win the needed margin of votes for victory; thus available funds are recognized as a major determinant of the success of individual candidates, as well as of parties.

Coincidentally, we seem to see a notable rise in the proportion of wealthy candidates, while the public and the legislatures and many public figures debate the impact of this correlation upon the public weal. Because a personality who comes across effectively on the TV screen may have far more appeal than the issues would warrant, appeals to the emotions, to prejudice and bigotry, have been provided with an unprecedented platform—a terrifying advantage for the rabblerouser and the demagogue.

But there are concurrent effects that are even more profound than those of the vast mechanical proliferation of political manifestations. And these have to do with the basic underlying values—values that preceded systems and that may well have more lasting importance than the systems that sometimes tend to obscure and neglect them. These values apply both to personality/character and to principles.

In this Age of Systems, candidates are evaluated as television personalities; before the Age of Systems their basic characters may have had greater importance. In this Age of Systems, candidates are guided in the utilization of mass media to attain concrete purposes by hired professional experts; before the Age of Systems candidates were guided (if at all) by like-minded friends and colleagues.

In this Age of Systems, all sorts of scientifically based techniques are brought to bear to learn the current trends of popular thought, to measure the relative proportions of the public holding various views, to test the impact of ideas, proposals, even modes of expressing points of view. Before the Age of Systems, candidates generally were clearly identified with party affiliation and platforms and often with decisive stands on controversial issues. Respected candidates—William Jennings Bryan, Norman Thomas, and others—ran, election after election, on virtually the same issues, defeat after defeat, in defense of principles they held dear; and loyal conscientious minorities were glad to vote for them even without expectation of success.

The *New York Times* recently carried a front-page article headlined "Computers Counseling Candidates."[1] The item said that "a governor up for re-election" asked a public opinion analyst organization to measure voter reactions to "his strong public stand on a controversial tax issue." The governor's stand was fed into a simulation program which made use of a representative sampling of voter attitudes.

The word came back that if the governor wished to attract "the younger and relatively liberal Republicans he was trying to woo," he should reverse his position. "Five days after receiving the report, the Governor reversed himself on the issue." So much for moral commitment.

Further, reported the *Times,* "This is only one of many examples this fall in which political technology, a science only dimly perceived 10 years ago, is emerging as a major factor in the 1970 elections. Increasingly, candidates are coming to rely on its techniques for guidance, inspiration and, in some cases, preservation."

However, it is recorded in Proverbs that "Where there is no vision, the people perish." In this Age of Sys-

[1] *New York Times,* October 30, 1970.

tems, man can look ahead with a well-founded prescience never before open to him. He can know more, analyze better, and evaluate more accurately. He can make wiser and greater plans, and execute them more efficiently. But none of these emerging capabilities rests upon the spontaneous reactions of the relatively uninformed. If the vast benefits attainable are to be realized, some leaders of vision must also have the courage and the strength and the will (and perhaps other nonsystematic characteristics) to address themselves to the long-term public good, rather than to the instant public reflex.

The Age of Systems has demonstrated its potential for the identification and the exploitation of political opportunities. It is now the responsibility of those who would lead to show how this capability can be developed to facilitate wiser decisions on the basis of the principles involved, rather than to serve mainly as a tool for well-financed opportunism.

5 The Organization and the System

ORGANIZATIONAL centralization (in some degree) is essential to the process of systematization. The conception, planning, designing, and coordinating of the system can scarcely be shared among all the personnel involved. If a central managerial or planning core does not exist for such purposes, but an ad hoc group is selected and formed to develop a system or to increase systematization, then that group must be regarded to all intents and purposes as central.

The most common characteristic of physical systems seems to be centralized control (that is, central to the system), subject to centralized management (that is, central to the organization). Many systems come into being only when—and because—the means for centralized control become available to management. The many forms of transportation systems present prototypical examples: Without central scheduling and dispatching there would scarcely be a system; and without means of control, such as the reporting back of unit locations, the systems could scarcely function on a modern basis. Manufacturing operations typically have centralized control—the more systematized, the more centralized. The control console is an archindicator of the systematized operation.

Centralization

Individuals working in systems typically think of themselves in relation to a central organizational nerve system, from which the ultimate authority emanates. Even in the so-called decentralized organizations, where considerable authority is granted to the local manager, his subordinates have no difficulty in understanding that he represents a central authority, and they often talk and think of this remote entity in terms of its location: "Wonder what New York will do about it?" or "Do you suppose Chicago knows about this?" The executive suite is the physical symbol as well as the actual seat of centralized authority in the organization.

The looseness or tightness of the organization does not alter the need for and the fact of centralization, but the looseness or tightness may affect the kind and degree of systematization. For instance, some retail organizations operate on a nationwide basis, utilizing mostly part-time employees or representatives (for example, those making house-to-house calls in their own neighborhoods). The organization can exert little or no direct control over these individuals, but it can systematize all aspects of the relationship in order to minimize the requirements for their direct contact with the full-time managers and to facilitate all communications and operations. And it can insist on compliance with system requirements. If sufficient incentives are available, this organizational arrangement can sustain a viable organization, for the intended purpose.

By contrast, an airline, with tens of thousands of employees and hundreds of millions of dollars of installations and equipment, requires the utmost in computer-assisted systematization in order to handle and control, accurately and rapidly, the many thousands of different individual transactions and operations that go to make up a day's business. Employees are extensively trained and rigidly supervised to insure maximum compliance with system requirements. Each operation is elaborately standardized and must be performed in a specified way. Uniforms and badges symbolize absorption into the system. Employees of the organization are acutely conscious of their involvement in the system and usually of the interrelationship of their functions with the functions of others in the organization. System and organization approach the stage of being coextensive—all but identical.

Some systems, by their very nature, tend to bottleneck when they reach a self-defeating size (like the dinosaur?). An interesting example is the assignment of direct-dialing telephone numbers to individuals when the organization outgrows the efficiency levels of a wholly centralized switchboard. Many systems are, of course, subject to overload effects; and demand factors, for instance, are not always controllable (as in extreme peaks of power demand, water supply shortages, and others).

Such conditions often require the addition of decentralized facilities to relieve overstrain on the focuses of the system, and much planning is devoted to estimating the provisions that should be made for flexibility within the system or for available supplementation—all at some sacrifice of the level of efficiency attainable on a pure system basis. The great development of services offering temporary employees is one indication of the degree to which organizations increasingly rely on systems of fixed or limited capacity, and are prepared to rely on outside agencies for any necessary supplementation.

Systems lend themselves to and usually, in fact, require centralized control. An organization that is large and diverse usually operates a number of systems, and the information derived from the operation of each of these can usually be transmitted to the central management of the organization with relative ease. This is often done via one or more intermediaries, such as department heads. Apart from operational efficiencies, then, systems offer certain inherent features attractive if not essential to centralized management.

There is, of course, a difficulty or deficiency inherent in such situations, as in any other where information is abstracted, refined, summarized, condensed, and otherwise made digestible for those at higher organizational levels. Something is left out.

Differentiation and Specialization

A primary characteristic of systems is the breakdown of operations into segments or phases, each of which can be handled on a concentrated, standardized, methodical basis. The operations within each separate segment or phase of the system are as uniform as possible; the differences exist among the segments or phases.

Where a segment or phase requires the services of more than one employee, their work assignments are likely to be identical or to vary only incidentally to the division of the work load. Thus clerical employees may "divide up the alphabet" or otherwise find the total volume apportioned among them, but in most cases this becomes a distinction without a difference because the units are similar and are processed similarly. Thus it is a basic characteristic of systems that the operations performed by individuals at the same stage are relatively undifferentiated.

By the same token, there are distinct differences among stages of a system if any significant progress is to be made in the processes involved. Each stage except the first receives its main input from a preceding stage, and its output constitutes the primary input of the next stage. Thus each stage is specialized.

Employees operating at any stage may detect certain evident deficiencies or anomalies—deviations from standard—in the items of input received, but they cannot always determine if such deviations originated in the immediately preceding stage or otherwise. However, it is exceedingly common in many systems that substantive errors of input or errors of processing are not detectable, or not readily detectable, in subsequent stages.

Thus deviations of form may be evident, while deviations of substance may be undetectable, at subsequent stages. However, the probability of unacceptable deviations of form is not great, because these are even more readily detectable at the stage where they originate. The errors of substance constitute a far more serious matter and, generally, present much more difficult problems. These are, to a considerable degree, inherent in and fostered by the essential differentiation and specialization of function from stage to stage of a system.

The necessity for differentiation is obvious, and the potential advantages of specialization can be great. But in many situations specialization is a matter of only superficial benefit and may lead to serious deficiencies. For instance, in many clerical systems the degree of skill or the extent of knowledge required for any one operation may not be very great, and yet the absence of one or a few individuals assigned to such operations may bottleneck the whole system and leave employees in subsequent operations with nothing to do.

Many managers recognize the functional advantages of having some employees trained to perform two or more tasks, and they plan deliberately to provide for absences due to illness or vacations in this manner. Others rely on foremen, supervisors, managers, or other graduates of the task to leave their higher paid and usually more important assignments to fill in at the less prestigious post when this sacrifice is necessary to keep the system going. (We are all too familiar with the examples of strike-bound operations maintained by supervisory personnel.)

When the various tasks occasioned by the planned differentiation and specialization of the system design are occupationally related and graduated in difficulty, an obvious logic emerges for initial and subsequent assignments; and perhaps even, to some extent, for career development as a functional program. Newly recruited individuals are placed in a position that requires minimum training; they are moved to the next easiest position and so on as they qualify and as vacancies occur. This provides a degree of flexibility in the overall staffing of the system, as each employee can presumably be called on to substitute in each position he has previously filled.

Such situations often give rise to serious abuses, however. The needs of the system are paramount; considerations involving individuals are subordinated; thus individuals are called upon to meet such demands of the moment as they may be qualified to handle. This can result in a doubling of the individual's work load, as he necessarily performs his regularly assigned task and still has to cover for an absentee. Yet an acceptable alternative may not be available to management, if relatively few individuals in proportion to need are qualified for each task.

This situation is particularly likely to occur in systems that operate on a basis that requires, for instance, daily completion of tasks, such as many accounting operations, where each business day brings its own demands for completion of records. But when some lag in task performance is permissible, the accumulating backlog can assume appalling dimensions—enough to intimidate an employee who finds it awaiting his return from absence due to illness or vacation.

Such problem situations as these, related to and caused at least in part by the differentiation and specialization of task essential to the functioning of systems, can be alleviated only by a departure from the rigidity of system prescription. Some flexibility—some provision for adaptability and adjustment—must parallel the strict, pure functioning of the unrelieved system.

Hierarchy and Succession

It is a truism that the personality characteristics most conducive to achieving success in large organizations differ from those that favor attaining success in more individualistic endeavors. It is also widely recognized that the characteristics most likely to speed the new young employee's progress through the lower grades differ from those more valued at higher levels—and even more from some of those probably essential at levels where important decisions are made.

These differences are greatly emphasized in most situations where systems are largely involved, and they have introduced a factor into organizational development that has created far more problems than are generally recognized—far less met, and solved. Widespread failure to cope fully with these problems has introduced or enhanced serious sociopsychological conflicts, the consequences of which are at times all too evident, but inadequately evaluated.

The psychological profile of the ideal employee of most highly developed systems (especially in paperwork) would include characteristics conducive to satisfactory adjustment to routine, repetitive, narrowly defined tasks that provide little opportunity for the attainment of personal satisfaction through individual achievement. There would be a minimum of the characteristics that would militate against such adjustment; ambition, drives toward self-actualization (in the sense employed by Maslow), and a need for meaningfulness in one's work would be minimal.

The psychological profile of the ideal supervisor for a highly developed system would include a somewhat different set of plus and minus characteristics. He must be capable of living simultaneously with a large number of administrative problems, some urgent, all differing in detail. He must also possess knowledge, competence, and a sense of responsibility broad enough to embrace the totality of the system and all phases of its operation. He must know how to exercise the authority vested in him to produce the necessary results with a minimum of undesirable effects; and he must have the drive and the initiative to devote himself as fully as necessary to make the system work well.

The difficulty lies in the simple fact that these two psychological profiles are far apart and "never the twain shall meet." In their prototypical forms, the ideal systems employee can never become the ideal systems supervisor.

This basic psychological reality is largely obscured by certain circumambient conditions:

1. *Numerical proportion.* There is such a large number of low-level employees per supervisor that the psychological eligibility of most of them to attain a supervisory level is scarcely considered.

2. *Turnover.* The quitting rate of low-level systems employees is generally extremely high, and it is probable that few of those with potential for more valuable contributions come to be regarded as permanent employees.

3. *Transfers.* The career-conscious employee looks about for a favorable environment and better chances; he will seek out other locations within the organization offering greater opportunity, and request transfer. If denied, he will probably quit.

4. *Palliatives.* There are usually minor posts within systems to which employees may be promoted on seniority or merit bases and which carry increases in pay and some improvement in status. These can often satisfy most of the ambitions and aspirations that exist among the generally modestly expectant low-level workers. In fact, if the post involves more exacting functions, many of those invited to consider it will turn it down, saying,

"Who wants headaches?" When this occurs, it offers a powerful confirmation of much of this thesis.

At the next major level of career transition, too great involvement with systems can again exert a seriously limiting effect. The administration of a highly developed and relatively rigid system, necessarily involving, as it usually must, a degree of specialization inappropriate to line operations or broad policy formulation, is seldom regarded as an excellent preparation for general management.

To the extent that these serious discrepancies are recognized, they are thought to be countered by management development programs, often largely based on college recruiting. But this has tended increasingly to create a two-class society within the organization, often with somewhat disruptive effects as nonmembers of the "jet set" are disgruntled by the contrast between the favoritism shown the newcomers and the disregard and neglect of themselves.

Where such rigid precedents of succession as strict seniority are followed, considerations related to systems tend to introduce factors that may affect the appropriateness of the appointment, sometimes making seniority irrelevant. On the other hand, when successors are chosen on merit, factors related to systems involvement (in former or prospective positions) may be extremely relevant but inadequately weighed.

From all these considerations, it is apparent that systems operating within an organization may often exert substantial and even inordinate effects upon the hierarchical evolution and upon the probabilities of succession. Many of these effects may be unrecognized as such, but their consequences to the organization can well be drastic.

Long-Term Effects

While it may seem absurd, an implicit question does exist in many situations where organizations focus upon and develop a high degree of dependency upon systems. That question is, Which has priority, the organization or the system? At first this question may not appear to be meaningful, but in many situations adequate consideration of the question may shed light on the obscure origins of serious difficulties.

It seems clear that it is the organization that has explicit and implicit goals, ends, objectives, and purposes. An organization is, by definition, a vehicle or means by which a number of people seek to attain collective objectives. The term *organization* is a collective noun, and it signifies a number of people in relationship to one another for more concerted action than would be achieved without such relationships. This multiple entity has, let us say, a goal. In a business enterprise the goal is basically to earn a profit by performance of specified activities calculated to bring about attainment of that goal.

The goal of a steel company, for instance, is to produce steel and sell it at a profit. All else is incidental—though perhaps indispensable—to achievement of the basic goal. Such a basic goal may be personally shared by relatively few of the people in the organization, perhaps only by the shareholders and members of top management. But the other members are induced to help in attaining that goal by offering them inducements—largely salaries and wages—which are actually derived from the goal-attainment efforts to produce earnings. This condition is clearest at the time of start-up when there are no organizational earnings and the incentives offered are all provided out of the capital supplied by the goal-oriented investors.

To attain the organizational goal, systems are set up and operated. In many cases these are considered to be essential to goal attainment. Great efforts are made to have the systems operate as effectively as possible. More and more attention is lavished on this specific objective, and this attention and these efforts tend to become more and more specialized. Eventually a point is reached where what is good for the system is not good for the organization. If the system-supporting endeavors are carried beyond this point, serious and perhaps even fatal damage may be done to the organization.

An example is to be found in some large "paperwork factories"—banks, insurance companies, and other organizations operating extensive and pervasive systems of data processing and information handling. The necessity for keeping such systems functioning with a full complement requires the constant efforts of large staffs for personnel administration—recruiting, interviewing, hiring, assigning, rating, compensating. When shortages of suitable job seekers occur, the pressure is on these staffs to fill the vacancies, and they are often compelled to sacrifice standards to expediency. Individuals are hired who would not be hired under other conditions. Individuals are placed in positions according to the needs of the system, disregarding the characteristics, special capabilities, and interests of the individuals. Supervisors and managers concentrate on employee's performance in present assignment and tend to ignore potential for other, perhaps more satisfying, work. Individual employees become aware that the real concern is for the system, and that concern for the individual is superseded. Boredom and frustration with stultifying routine are unalleviated by the expectation or hope of more rewarding work or by sympathetic relations with superiors.

The brighter, more ambitious, less patient employees leave, and they take with them much of the organization's future potential. The employees who stay create an atmosphere of sullen minicompliance; they do the least they can to keep their jobs because doing more is unrewarded. Errors proliferate, sloppiness spreads, supervisory and managerial problems increase. The functioning of the system is impaired and in some cases there is an actual breakdown.

Too intense concentration on the system has been self-defeating. It has disregarded the broader needs of the organization, and it is for the attainment of organizational goals, after all, that the system was called into being. Which should have priority?

The answer, perhaps, is that they should not be viewed separately: Whenever a decision is to be made to benefit the system, care should be taken that it does not affect adversely other values of the organization.

Can the organization exist without the system? In some cases yes, but with adjustment and adaptation. In other cases no, as it would not be economically viable.

Can the system exist without the organization? A system may exist in a changed or different organization, but it can only exist in an organizational environment.

Thus what affects one must affect the other. And those who make the policies and those responsible for carrying them out should develop a broader view than that concentrated on the operation of the system alone. The welfare of the host organization should never be the victim of short- or long-term sacrifice, even if unintended.

Regimentation

The military have always emphasized systematization. The requirements for managing the performance of large numbers of men made coordination essential. Training in standardized patterns and rigid discipline were essential. Logistics and transportation required methods suited to available equipment and supplies; siege, attack, defense, and pursuit all required tactics fitted to available weaponry and other significant factors.

The horse, the chariot, the spear, the bow, the catapult, and the cannon all in turn offered opportunities for the introduction of new systems, as did the machine gun, the tank, gas, the panzer column, the fighter plane, the bomber, the parachute, and other products of technology adapted to military use. In every case, the generals who created the most effective systems around the new weapons won successes that unsystematic use of the weapons alone could never have brought them. Perhaps the most striking example of this was seen in the tank battles in North Africa in World War II, when the newest model of tank often seemed to have as much effect upon the tide of battle as the skills of General Erwin Rommel and of his opponents.

From the phalanx of Alexander to the hollow squares of Waterloo, from the elephants of Hannibal to the longbows of Agincourt, from the Roman roads of the Caesars to the cavalry of Lee, the creative endeavors of military leaders sought combat effectiveness that would not depend solely on the prowess of the individual warrior but rather on the total system of warfare within which the effectiveness of the individual would find systematic expression.

But the traditions of soldiering existed long before the Age of Systems. The soldiers of Alexander were largely Macedonian serfs. The English infantry at Waterloo were the sons of ignorant peasants and illiterate proletarians. The followers of Hannibal were largely Phoenician slaves. The longbow men at Agincourt were serfs and peasants. The Roman legionnaires were mostly ignorant mercenaries. Many of Lee's men were illiterate "poor whites." The equipment was simple; and the systems were and had to be simple because the manpower was unsophisticated.

The men, however, were accustomed to regimentation, and the hardships of the military campaign were little worse than those of the daily life at home—with no energy but muscle, and a standard of living commonly near the subsistence level. With looting privileges and the division of booty, if not pay, the economics often seemed attractive.

With the mechanization of warfare by advanced industrial nations, factors have been introduced that bring men to resent and resist some aspects of military systematization. Men who have been taught from infancy (or who have learned since) to expect and demand respect as individuals will no longer passively accept the assumption that a superior is always right, that discipline must be accepted, or even that all commands must be obeyed. The more comfort and social warmth the individual has enjoyed, the more he may resent—and question the need and the justification for—deprivation and hardship, extreme exertion, and the rigidities of organizational subordination.

Where there is scope for individualism, some individuals may find successful adjustment, such as the extensively trained and highly skilled fighter pilot, the accomplished mechanic, even the nearly autonomous master sergeant. But the vast majority, submerged in the great anonymous mass of the squad of the platoon of the company of the battalion of the regiment of the division of the army, inevitably must feel their individualities disregarded and their personal tastes and preferences violated. In addition, those who have become conscious of the difference between interest and tedium will find themselves, most of the time, extremely bored.

What are the inevitable consequences of such maladjustment? Alienation, first, from all that is associated with or appears to represent the system that oppresses them, followed by antimilitarism, insubordinate tendencies, bitter criticism of all aspects of the situation—including the governmental policies and national objectives and interests that brought about their involvement.

A reporter wrote from Saigon: "Many [soldiers] . . . wear peace bands, show irreverence if not disobedience to authority, smoke marijuana, avoid salutes, object to the war they are in and tell all about it to those who will listen."[1]

The drive to withdraw from the military system environment may lead to desertion or overleave, but is certain to find expression in opposition to conscription and perhaps in questionable acts in uniform, when opportunity or provocation occurs, from black marketing to

[1] *New York Times,* October 22, 1970.

violence and destruction, as displaced manifestations of resentment.

The temptations to compensate themselves for their sufferings will lead men not only to undue pursuit of relief and recreation but also to extreme forms of self-indulgence. Men will drink more heavily than they would otherwise. Some, who would never do so in civilian life, will visit brothels. And not a few will follow the drug culture that seems to represent the ultimate rejection of the Age of Systems.

The *New York Times* reported from Saigon: "Most soldiers questioned contend they use marijuana out of boredom . . . and many say they have such easy, routine jobs that they can work just as well 'stoned' as not."[2]

Sometimes the forbidden is defiantly sought for special objectives within the system. The article said: "One soldier . . . said he smoked marijuana constantly while on patrol. . . . 'and if enough people think I'm weird, maybe I can get transferred out of the field—which is what I am shooting for.' "

The alienation of some young men forced into the military is so extreme that it creates an unbridgeable gap between them and the career army men. The newspaper reported: "One young soldier, voicing the draftee's contempt for career soldiers, said: 'It's a lifer-antilifer thing.' "

The lifer accepts the system. The draftee resents and resists it, thus becoming an antilifer. The contrast was pointed up in the article: "For the career man, service in Vietnam is vital to his future. For large numbers of draftees . . . it is a waste of time. . . . They feel they

[2] *New York Times,* September 2, 1970.

are the victims of the Army, and they say that nothing the Army has done in the last 30 years is of interest to them. . . . As one cynic put it, the U.S. is really withdrawing from Vietnam because the lifers can no longer stand the draftees."

It should be noted in this connection that the military problems of the army in Vietnam (not those caused solely by its own personnel) have been served up most deliberately by an enemy that operates, for the most part, without anything that our military would call a system, but rather with comparatively primitive systems that appear to ask or demand a great deal of those involved in them, including extreme regimentation.

In the meantime, the army professionals constantly elaborate the system. Drew Middleton wrote:

> An electronic battlefield where sensors and 'people-sniffers' pick up hostile movement, infra-red beams guide infantry to the enemy and computers evaluate combat information, is the goal of an extensive research program. . . . The over-all practical objective is an Integrated Battlefield Control System that will take over most of the battlefield intelligence functions served from the earliest times by the patrol. This system's development from present tests, some staff officers predict, will force a reorganization of the field staff system introduced into the army by Baron Frederick William von Steuben in the Revolutionary War.[3]

The navy and air force, of course, are at least keeping pace.

For the foreseeable future, a great power must, ap-

[3] *New York Times,* October 27, 1970.

parently, maintain appropriate military capabilities. These will increasingly involve more and more technologically advanced systems. But they will also involve people, upon whom the effectiveness of the systems must depend. The effectiveness of the people should—must—be a prime consideration.

Fleet Admiral Chester Nimitz once held up a simple, common electric call-bell push button. "This," the admiral declared, "is the only part of push-button warfare that has been invented."

The day may come when nations can win wars by pushing buttons. (Now this can only mean mutual obliteration by the superpowers.) And the day may come when there will be no more war or threat of war. But until one or the other of these long-awaited days arrives, it is necessary that the military leadership bear in mind that no system can win a war without some individual participation; and that it should plan and provide for the contingent utilization of inconceivably costly equipment with this inescapable fact in mind.

Shortly after the holocaust of Pearl Harbor thrust the U.S. into World War II, the writer was directed to conduct an investigation at the Naval Receiving Barracks in New York City. The brig was full of prisoners charged with gross or repeated overleave or desertion; these men had all missed ship. Some senior officers ascribed this to cowardice, suggesting the men wanted to avoid sailing seas now made hostile by German submarines and raiders and Japanese fleets.

The investigation did not confirm a single case of cowardice. In fact, the analysis of virtually every case led back to personal frustrations, mainly due to unnecessary and remediable maladjustments that had been aggravated by the imposition of rigorous wartime conditions and the withholding of such alleviations as liberty and leave. The writer's subsequent experiences at sea reinforced the conviction that, when young sailors were helped to make acceptable adjustments to their basically unsought situations, they not only performed routine duties adequately, but also revealed unguessed resourcefulness, initiative, and creative improvisation, as well as a determined cheerfulness that often provided an invaluable feedback upward.

The military authorities have had to take a good, hard look at the problems of putting and keeping millions of generally reluctant young Americans in uniform and converting them into effective fighting men. And they have begun to do this by a growing recognition of the simple fact that each one is an individual—not merely an enlisted man—and each sees himself as himself, not merely in the role in which the service has cast him.

The armed forces are trying. They have recognized the value of giving the individual the privilege of choosing the specialization he prefers and of offering the opportunity to learn and advance. Unfortunately, there are inherent limits to such permissiveness. The needs of the service determine the governing parameters, but within these there is scope for a considerable degree of flexibility. In general, the services have demonstrated a disposition to extend the resultant opportunities to their personnel, though undoubtedly the administration has fallen short of the intent.

Most important, newer concepts of leadership, drawing from the social sciences, have been adopted for the training of officers, and newer principles and practices have improved the quality and spirit of military justice.

As technology increases the complexity of many military tasks, the requirements for knowledge, skill, and intelligence rise, and the continuous upgrading of personnel (in some parts of the services) tends to emphasize the individual rather than the all-but-indistinguishable "swabbie" of yore, whose capabilities above the most primitive were a matter of indifference, if not of suspicion, to his superiors.

Most hopeful of all is the growth of understanding among leaders of what it takes to lead—successfully and effectively—commands largely composed of men who are generally fairly well educated and who are, by and large, accustomed to some degree of indulgence of their personal tastes, interests, and proclivities.

A hopeful example was presented in a description of Admiral Elmo Zumwalt, chief of Naval Operations:

> His head tilted to catch each word, as some 1000 sailors met with him to sound off their gripes—some general, some highly personal—about military life. Quietly and sympathetically [he] responded to each. . . . Zumwalt is proving unusually well equipped in both inclination and experience to . . . retain and attract more volunteers at a time of widespread youthful antagonism toward the military. . . . Of all the suggestions he has heard, he has so far circulated more than 800 . . . for reaction from selected commands. Of these 65% have been turned into direct orders . . . to effect changes . . . aimed mainly at eliminating many seemingly minor, but unsettling, irritations.[4]

The magazine reported that a petty officer said, "Thank you, admiral, for treating us like people," and

[4] *Time,* November 9, 1970.

spoke of the admiral's "disdain for those traditions that demean low-ranking personnel."

We may well suppose that, if the quality of life can be improved by the interested, empathic initiative of military leaders, this would be all the more feasible for managers of civilian organizations, who are less burdened by regulation, tradition, and the need to uphold discipline; and ultimately less responsible for such stark issues as life and death, victory or defeat, national interests, and, perhaps, national survival.

6 The Individual and the System

THE word *individual* signifies a unique entity; every human differs from every other human. These differences may be insignificant to some, who value or who are concerned with the individual only in relation to some limited criterion or purpose. But these differences exist, and they provide the basis for the subjective awareness of differing from others which, however vaguely or inaccurately, affects in some way the social outlook and the behavior of everyone.

Every individual has some sense of his own qualifications to perform—or learn to perform—tasks of all kinds. This sense varies from individual to individual and, from moment to moment, within the individual; from a defeatist certainty of failure to a greatly overconfident self-assurance. However well- or ill-adjusted to realities, this sense largely governs, in the long run, the individual's acceptance of and adjustment to his occupational employment.

Personal Qualifications and Characteristics

Each individual's value system reacts differently to the forms and conditions of employment. Some are most affected by more or less abstract considerations, such as those seen as related or contributing to some kind of status. Others find adjustment most readily when the economic return meets or approaches some subjective standard; they appear to be crassly materialistic in their evaluation of their jobs. For most, however, many other factors enter—environmental, physical, social—including the actual requirements of the job in terms of the specific behavior with which it is assumed they will fulfill the expectations of their superiors.

As a matter of common observation, the attitudes of individuals to their employment seem to account for the most extreme differences in the value and nature of each individual's contribution. This can range from the most dedicated devotion to meeting organizational expectations, arising from what is apparently complete identification with organizational goals, to disinterest, alienation, and hostility—expressed in minimum productivity, "testing the limits" with superiors, constant expressions of dissatisfaction, and demoralizing fellow workers.

It should also be noted that the most apparently loyal employee may be primarily motivated by sheer insecurity, while his potential contribution is limited by his marginal qualifications. On the other hand, the troublesome employee may be extremely competent, and even possess great, if uncalled for, potential for higher levels of employment.

The impossibility of creating or transferring innate capabilities is generally recognized, but the transformation of attitudes is quite generally regarded as possible.

Thus a superior would recognize the futility of attempting to make Mr. Loyal as intelligent and basically capable as Mr. Disruptive, but he might wonder how he could inculcate in Mr. Disruptive some of the cooperative and generally supportive attitudes of Mr. Loyal.

When we approach the problem of a manager considering the relation between an employee's qualifications and a specific position, we appear to be concerned with a number of not easily determined quantities. Some of these appear to be:

1. The specific requirements of the position.
2. The specific capabilities of the employee, in relation to the requirements of the position, based on: experience, present or future; training, if experience is inadequate; and potentials, if experience and training are inadequate.
3. The degree to which the employee will actually apply his capabilities to the position.
4. The excess potentials of the individual applicable to a more advanced position for which the position under consideration may be preparatory.

It is rare for a position to be created especially for a certain individual, but it cannot be as rare as the creation of an individual specifically for a certain position! Thus matching individual with job can probably never be an exact science. But surely it can be more exact than simply to judge whether the capabilities of an employee meet or exceed the requirements of a position.

While the excess potentials of the individual are sometimes considered to some degree (not as often or as reliably as personnel appraisals may suggest), their greatest immediate—and ultimate—importance will depend on their effect on the application of capabilities by the employee.

In this connection it seems desirable to note that the manager usually cannot in any way affect the basic entities represented by the requirements of the position or the capabilities of the employee. He may be able to affect the degree to which the employee will apply his capabilities.

This leaves the specific requirements of the job as fixed and unalterable, a standard against which all else is measured. Is this essential? Is this inevitable? Is this immediately or ultimately desirable?

In smaller and more loosely controlled organizations, managers and supervisors can and often do adjust the assignment of duties to employees. This can be done in accordance with individual characteristics, capabilities, interests, and attitudes. It need not always be done by seniority, or in relation to the existing pattern of work distribution, or in accordance with other considerations that may be, in principle, largely irrelevant, however plausible and obvious they may appear as guides to administrative decisions.

But where systems are installed, the discretionary scope of the manager shrinks or disappears. The job requirements are fixed. Recruitment, training, and assignment are concentrated on specific capabilities of the employee. Excess potentials of individuals tend to be largely ignored; and the application of capabilities by employees suffers, usually to the point of minimal conformity with standards and with growing requirements for disciplinary indulgence.

The job may ask little—often too little—but it asks that little uncompromisingly. And too often it offers almost no reward in terms of personal satisfaction earned by adequate performance. Rewards and compensation are all extrinsic as there is little or nothing intrinsically satisfying in the task. William Braden wrote: "The instrumentalization of things [is leading to] the instrumentalization of man."[1]

An individual who is goal-oriented—imbued with a definite purpose—can more easily endure even those tasks obnoxious to him but will perform consistently well when they are accepted as incidental or instrumental to the pursuit of his objective.

Paradoxically, however, the goal-oriented individual will not stay in an unrewarding situation any longer than he feels justified to serve his long-term purpose. Thus the system operation that requires workers for dull, routine, uninstructive tasks with no likely future is doubly frustrated: Those capable of tackling the work without apathy will leave, and those who are willing to stay are most likely to be or become apathetic.

Some systems thus present, in the persons of a part of their personnel, a situation involving substantial entropy.

Effects of Systems

When we speak of participating in or becoming an integral component of a system, let us keep in mind significant differences between the activities required of an industrial worker and those required of a clerical worker.

[1] *The Age of Aquarius: Technology and the Cultural Revolution* (Chicago: Quadrangle Books, Inc., 1970).

The industrial assembly line worker is involved in nonverbal activity and deals—repetitively and monotonously—with material *things,* with *reality,* with materials and with machines; with parts and with processes; with components and ingredients, equipment and tools. These may engage some degree of his attention (less and less with simplification and repetition) but the required activity does not engage the higher levels of the intelligence. The mental input for his function tends to be relegated to rigid, regular, habituated patterns of activity. Since all the units of reality that appear before him to receive his repetitive contribution are supposed to be exactly alike, he need be alert only to differences. After a few days of habituation, a mental "set" is arrived at that can be triggered by a deviation, but that requires little or no conscious attention. Thus even self-inspection usually fails to engage the higher centers of the mind.

But the clerical employee—if actually performing clerical work—faces an entirely different class of requirements. He is in a basically different kind of activity, requiring a different degree of actual mental participation in the work. The clerical employee, by definition, deals with symbols—with letters, numbers, words, meanings, and data. Where symbolic processes are involved, the attention required levies upon higher levels of the intelligence; anything less than the required degree of concentration must result in mistakes—error—frustration of the system itself. The monotony of many clerical tasks derives from their repetitiveness, but the nature of the repetition itself differs from that which is characteristic of industrial-type repetition. For instance, one common, low-grade clerical operation nowadays calls for the clerk to read a name (or other symbolic reference) on a paper (form, letter, list), and then find a prepunched card bearing that reference in a large file containing many such cards. (The paper and the card are passed on to a key punch operator, who punches additional data from the paper into the card.)

This operation is about as simple as any clerical operation can be. And it is crushingly repetitive and monotonous. Nevertheless, each individual operation *is* inherently different. Each paper bears different symbols, calling for different cards from different parts of the file. The clerk must translate the relevant symbols into behavior that is sufficiently precise to insure that the correct card will be selected from among the hundreds, or even thousands, of possibilities in the file.

The symbolic-interpretive levels of the clerk's mind are constantly engaged, although the content of the symbols is otherwise meaningless to him. So far as his interests or values are concerned, one name is like another. There is no possibility of attaching other than immediately operational significance to any of the symbols involved. There is familiarity, but no true recognition.

The job is a menial one, as clerical jobs go. Anyone who can read can be trained in a few hours to do the work. Within the week it will be utterly monotonous. It is, perhaps, somewhat typical of the involvement of many individuals in clerical systems.

But what of the quality of such work? The requirements may be low, but what are the opportunities for deviation, for error? What could be the consequences of misreading a symbol on the paper, or picking the wrong card?

"Importance" has many dimensions. How important is error-free performance on this menial task? In most such situations, there is no correlation between the status of the task and the dimensions of the consequences of error. An error that originates in such an operation may be extremely costly or damaging. But this fact generally does little or nothing to inspire responsibility when the work is so undemanding as to place the worker at the bottom of the system's status ladder.

Clerical activities may be classified in many ways. One useful dichotomy divides clerical work into activities involving public contact and those involving only contacts within the organization. Typical of the first are the bank teller, the airline reservation clerk, the sales clerk, the cashier of a pay-as-you-leave establishment, and the receptionist. Each of these operates within a narrowly prescribed range of procedures, varied only in a way that was generally foreseen in their training.

Each contact is different; but the exception is rare, and usually requires referral to or calls for assistance from others. Within the range of variations with which the clerk is qualified to deal, there is a consistency of required behavior that amounts to a pattern, a generalized conformity to authorized procedures that, despite the variation in details, invites monotony. In such positions, however, this monotony can be relieved to some extent by the informal, nonfunctional, conversational aspects of the individual encounters. This relief is often limited, however, by the pressure of work, such as a waiting line of customers, or by the formalization of the interchanges imposed by the organization itself, as with telephone company operators.

The principle illustrated here applies within an organization, too. Some clerical positions involve contact with others, perhaps even contact with other departments of the organization. Too many clerical assignments, however, require the clerk to sit at a desk all day, with no task-involved contact during working hours. Many employees find it difficult to tolerate such minimization of interaction, especially when occupied by a monotonous task. Coffee-break and lunch-hour socialization becomes a virtual necessity for these employees under these conditions.

Another basis for classifying clerical tasks is related to the opportunities they provide for learning of and understanding significant operations of upper-level individuals or of parts of the organization or of the organization itself. A classic example is that of the secretary who knows what is going on and feels herself to be a part of it all. She can tolerate the tedious filing and the stenography that are part of her job if she finds enough intangible satisfaction in the aspects of her work that are conducive to a sense of participation in matters of genuine interest within the organizational environment.

Senior clerks are sometimes in a position to follow trends of major interest, to possess inside, confidential information, to know in advance of important developments, even to contribute to or participate in the bringing about of major results. Such clerical positions are rewarding to their occupants according to their individual dispositions, and the sense of participation, of being "inside," can compensate for much monotony in the actual work involved.

Still another basic classification of clerical tasks is that concerned with the breadth or "enlargement" of the required activities. A clerk who conducts, himself, all phases of a clerical process not only has variety in his work but also has the opportunity to know and understand the reasons for and the results of his work, and to derive a sense of accomplishment from it. Balancing an account is one example of this, whether it is a stock-record clerk in a brokerage house balancing the firm's position in a security or a restaurant cashier balancing cash, checks, and credit charges against the total of the day's business. Unless there are undue frustrations, such activities can provide a certain inherent satisfaction to some individuals because of the sense of accomplishment involved.

If we now review the three bases of this classification, we find that contact with others, understanding and being interested in what the job is about, and having a sense of accomplishment in one's work are all conducive to satisfaction on the job; whereas minimum contact with others, ignorance of the ultimate purpose and value of the job, and finding one's work meaningless are all conditions that allow the inherent monotony of the tasks to become intolerable.

Here are three sets of conditions—three positives and three negatives. Two positives can generally outweigh one negative, but not always, as the negative may be too strong for the individual. And one powerful positive may sometimes balance two negatives, if these are moderate in intensity. But three negatives must overwhelmingly prevail.

And that is the situation that is most likely to overtake the individual clerk who is assigned a part in most large, tightly knit clerical systems. He finds himself compartmented, without understanding of or perspective in the system in which he functions, and without the gratification of sensed accomplishment. Yet he must face the daily grind of processing symbols, with utterly repetitive monotony, with no real meaning, and often under pressure for greater productivity. Behind all this, the system's requirements for accuracy impose a relentless burden of attention—of concentration on a task utterly without inherent satisfactions.

Inevitably, the individual will escape the intolerable by making a satisfactory, self-effacing adjustment, by quitting, or by compromise. The compromise usually adversely affects his contribution to the functioning of the system to such an extent that quitting is likely to be less harmful to the organization.

People who find themselves in an unsatisfactory situation usually look around for a way out. If they fail to escape and decide there is no use in struggling, they begin to adjust, to adapt, to rationalize. They have a psychological need to become reconciled to the inescapable.

Adjustment to an unsatisfactory situation is possible only by changing one's perception of it and of the factors that characterize it. Thus one tends to (1) emphasize whatever is good about the situation, (2) deemphasize whatever is bad about it, and (3) seek favorable comparisons with actual or potential alternatives or with the more or less parallel situations of others.

Thus a worker who is not really happy in his work can tell himself:

1. Well, at least I'm getting paid for it.
2. I'm getting used to it.
3. It could be worse.

The willingness of most American workers to adjust to their working environments has made it possible for American management to maintain these environments at their unnecessarily unsatisfactory level.

Depersonalization

Psychologists recognize, history teaches, and many perceptive observers have noted in particular situations that the forcing of numbers of individuals together under nearly identical conditions has a predictable effect upon them, even when resisted. Such conditions act upon them to negate their individuality. Extreme examples of this are seen in penal institutions, the military, and other situations where great uniformity of conditions and fixed requirements of behavior are imposed.

Participation in closely disciplined, identically performing groups is not always an unpleasant experience, and some persons have more appreciation of or tolerance for it than others. Additionally, many find some satisfaction in even the most severely regulated group activities, such as calisthenics, precision drilling and marching, ballet classes, and choral singing. There is a sense of group participation that provides a gratification not otherwise attainable; it is, presumably, an aspect of the sense of belonging, of group acceptance of each individual involved.

Groups, large and small, may be characterized by the shared attitude of their members—their esprit de corps; and this may develop into extremely high morale, as in some elite organizations where membership confers a sense of status and fulfillment, or an implication of having met certain standards, or of possessing certain qualities or attainments.

On the other hand, there are conditions under which group morale may be extremely low, and membership in such a group may be a frustrating experience. Forced membership in a low-status group must constitute a serious cause for dissatisfaction; and when activities are imposed that are not in themselves satisfying, the experience may be seriously disagreeable. In extreme forms, of course, such as certain prisons and prison camps, the unfortunate inmates are subjected to great suffering; and their physical distress is often matched by the psychological ordeal of total depersonalization, as when all are wearing similar garb, doing the same work, eating the same food, following the same dreary routine, and being distinguished from one another only by tattooed numbers.

Some of these depersonalizing effects tend to be characteristic of certain situations relating to the operation of systems, and especially to those forms of employment for which, typically, the recent high school graduate is eligible and sought. Although participation as an employee in the organization is certainly not compulsory, it is probably true that many of those involved are motivated to join only by a lack (or lack of knowledge) of available or preferable alternatives. This can scarcely be considered the most promising basis for recruitment, but it could be overcome if conditions provided a basis for the building of positive morale.

However, the sometimes humiliating routine processes of recruitment, hiring, and assignment are often as impersonal, and thus depersonalizing, as those of the system in which the employee is to work. The very introduction to the job may predispose him to sensitivity to the loss of significant personal identity. This more or less conscious anticipation is usually reinforced during his perfunctory indoctrination, his reception by a harassed supervisor, his casual turnover to another employee for training or his membership in a group being processed in a school, and his uneasy breaking in on the arbitrarily assigned job, where commonly he is little aware of the purpose or the greater significance of what he is called upon to do.

And this is merely preface. He is soon performing the repetitive routine of the assigned task in company with a group of others. The task is not difficult; in fact, once learned, it is easy; but it is utterly, crushingly boring. Once mastered, it demands only a part of the attention and little of the ability of those performing it.

Work is received from outside of or from a preceding phase of the system and is passed on to a succeeding phase. A supervisor presides over the activities of the group, and is presumably knowledgeable and a potential source of information about the overall system and the significance of the part played by the work of the group and of the individual. But he is far too busy correcting errors, handling anomalies, conducting necessary liaison with other units and with his superior, administering the unit, handling details of reporting and of personnel records, processing departing and incoming employees, training new employees, and learning and implementing the minutiae of changes in policies and directives. The new employee too often functions under and with co-workers little better informed than he.

The new employee feels neglected, forlorn, unimportant. And this sense of unimportance heightens and emphasizes the process of depersonalization. At first the new employee also feels insecure because of his unfamiliarity with the environment and the requirements, but he is soon reassured by his fellow workers, who freely express the negatives of their own work experience: "You don't have to do very much to get by," and "Don't knock yourself out!"

Other negatives current include "You probably won't

stay very long anyway" and "The supervisor is always busy—too busy to bother with us."

Under the pressure of such informal indoctrination, the new employee tends rapidly to conform to the norm of his group. He learns to perform more or less as required, quantitatively; but he also more or less disregards the need for accuracy as he learns either that responsibility for the errors cannot be pinpointed or that there is a substantial toleration of erroneous performance.

The new employee also learns that he will receive increases in pay and benefits after specified intervals of employment and that these "merit" raises are owed to the calendar rather than to any individual merit. All of this, with its obvious implication that individual performance generally goes unrecognized, unpunished, and unrewarded, exerts a powerful depersonalizing effect.

That such experiences may indeed produce drastic effects is reinforced by the views of such authorities as the psychologist Rollo May, who said: "No one who has worked with patients for a long period of time can fail to learn that the psychological and spiritual agony of depersonalization is harder to bear than physical pain."[2]

The many potent depersonalizing effects suggested tend to be focused upon the young, since new employees tend to be drawn from those currently entering the work force or those who have recently done so. This condition appears to be especially true of the larger management information systems, which generally require numbers of relatively low-paid workers, and can usually utilize unskilled, inexperienced young people by training them to perform simple, repetitive tasks.

[2] Rollo May, *Love and Will* (New York: W. W. Norton, 1967).

The young may have experienced some degree of depersonalization through participation with other young people in mass activities, and they may have adopted a substantial degree of conformity to youth norms in dress and in behavior patterns. But all this has been accepted and absorbed from a preferred environment and has been derived from peers.

It differs in the most fundamental way from the processes of depersonalization imposed by employment and by paid involvement in some routine task within a large impersonal system. The conditions and affective factors are radically different. There is all the difference between the satisfactions of belonging and the antipathies and alienations of exploitation.

And there is a high degree of correlation between the potentials of the young individuals for achievement and contribution with their susceptibility to adverse reaction and their resistance to and resentment of depersonalizing effects. The more intelligent, the more ambitious, the more energetic and conscious of capability the individual, the more certainly he will reject and rebel against an environment that refuses to acknowledge his special qualities and insists on viewing him impersonally as a kind of human module, quite like any other, to be considered only in terms of his matching the lowest common denominator required for a stultifying task.

In the light of the depressing and even frightening phenomenon of alienated youth that so engages the desperate attention of educators, parents, and governmental officials at all levels, is it not worth considering that, to so many high school or college graduates, the embrace of the establishment actually means the commitment of one's precious and potentially most enjoyable time of life to unacceptable conditions, unworthy tasks, and unedifying influences?

In all too many cases, the demoralizing effects of the terms and conditions of employment, taken as insulting and intolerable, are unalleviated by adequate efforts to excuse or explain them on any basis that might perhaps render them more acceptable to some on a temporary basis, as an essential phase of career development. It is not too strong a statement that the military conscript, who must suffer a far more severe period of preparation, is much more aware of its transitional nature.

Much light is shed on this vital problem by the comments of Jon McAtee, president of the student association at Northern Illinois University: "Students reject the 'role prescribed for them by society.' So they put up resistance . . . to social practices, such as conformity. When the university finds itself associated with the social pressures, it becomes a focus of resistance."[3] For *society*, read *system*. For *university*, read *employer organization*.

Downgrading Versus Upgrading

In the light of common experience, it seems we must accept the probability that each individual has a certain genetic or innate potential that cannot be extended by any measures currently available to the managers of private enterprises. Individuals perhaps never perform at the level of this theoretical potential, for a wide variety of reasons. These reasons run the gamut from lack of opportunity to lack of motivation, from lack of specific training or relevant experience to countervailing, preventive, disturbing, or distracting factors.

[3] *Time*, June 8, 1970.

When a machine is designed, the designers allocate to each part a specific function. No part is expected to perform beyond this specified function; indeed, excessive performance by a part could be disruptive to the whole. This is true even when the performance of the machine as a whole is improved; the part still must function precisely as intended in fixed relation to other parts and to the machine as a whole. If a part performs its function adequately, no other part similarly designed can be expected to perform better; any significant superiority of a part must be in terms of cost, endurance, smoothness, silence, maintenance requirements, or factors other than mechanical performance per se.

Humans, obviously, are different, not only from machine parts or even machines, but in their actual and potential performances. To no one should this be of greater interest than to managers, and, perhaps, to no managers more than to managers of systems.

When a tool or machine is applied to uses below its potential, there is an economic waste though the machine may last longer and require less power and maintenance. When an individual is called upon to perform, regularly and systematically, tasks far below his capabilities and is, in effect, denied the opportunity to utilize his capabilities, there is also an economic waste.

But there are other effects, because an organism is involved. Much as unused organs atrophy, unused capabilities decay, unused knowledge evaporates, unused experience is forgotten, unused skills disappear. There occurs a loss of individual capability and of the individual's potential.

But that is not all, and perhaps not even the worst of it. Behavior patterns influence subsequent behavior and have their effects upon the individual's value system, his perception of the outside world, and his concept of himself.

These factors can never be static, nor are they necessarily all operating in the same direction. But the resultant of all these factors—the totality of the vectors—will be generally positive or generally negative in its significant effect upon the individual.

Thus if an experience enlarges the individual's internal resources and facilitates the use of a greater part of the individual's potentials, it may be said to have an upgrading effect. But when an imposed behavioral pattern tends to reduce the individual's capabilities and makes less likely the fullest utilization of the individual's potentials, it may properly be said to have a downgrading effect. This destructive effect upon his own persona is a part of the price that an individual pays for submitting to employment that asks too little of him besides toleration of workplace conditions that provide little or no exercise of or stimulation to those elements of his being that represent his potential for significant, intrinsically satisfying, socially useful achievement.

No material compensation can adequately balance such a sacrifice, yet, paradoxically, the forms of employment most commonly involving these conditions are among the lowest paid. Surely the wages or salaries they offer constitute a bad bargain for some, if not many, of those who accept them, and the bargain is worst for those who have the most to lose.

In dealing with human values we involve ourselves in philosophy, perhaps even theology. Who is to say what is or is not a suitable occupation for an individual? Gandhi set the example of spinning cotton by hand;

Thoreau earned his daily bread as a gardener; Socrates dialogued while loafing; a vast proportion of the earth's population appears to find life tolerable and sometimes even enjoyable while manifesting little interest in developing its productivity beyond that required to provide the basic necessities; all sorts of human potential is devoted to the quest for knowledge that may never be put to use. When all such considerations are brought to bear, who dares assert arbitrarily that the good life must be measured against the standard of maximum use of maximum potential for economically significant productivity?

Although it is foolish to be arbitrary and it is cowardly to avoid the issue, it becomes necessary to seek a workable basis for proceeding, in the expectation or hope that it will prove reasonably acceptable to most of those who will utilize it.

Therefore, no implication is intended that a Hottentot or a fellah or a Pygmy is necessarily upgraded by being trained as a cook, a mason, or a mechanic, or even by being educated to practice a profession. Far less is it suggested that a chairman of the board or a vice-president—marketing is necessarily downgraded when he resigns or retires to teach or write music or play golf or fish.

When a farm worker becomes a mechanic or a truck driver, his earnings are increased and his standard of living can rise. If he dislikes the change and returns to milking cows, he must usually sacrifice some of the amenities and benefits. Which change represents upgrading and which downgrading is a matter of the individual's own judgment or opinion or feeling.

The individual directly involved is the logical one to evaluate his own condition, because only he can weigh the intangible, largely inexpressible values, mental associations, preferences, aspirations, hopes, expectations, and other phenomena of his inner life. Others, of course, can and will form their own views, but these must necessarily rest upon assumptions, supposed insights, imperfect observation, and incomplete knowledge of the total situation, all affected by personal values and prejudices differing in many ways from those of the individual involved.

The individual has his own basis for knowing or believing that he is being upgraded or downgraded. He may change his mind about this later, as different values become paramount, but he will have his own sense of the overall value, to him, of his occupational situation.

An experienced person who changes jobs voluntarily after substantial experience in his field intends to upgrade himself, but he may see this in terms of increased income, increased status, increased opportunity to advance, increased opportunities for achievement, or in terms of the specifically required activities that will occupy his time and attention. All these and more are combined, undoubtedly, though the proportions and emphases of the mix of evaluational criteria will vary with each individual.

Thus an experienced employee who moves by choice from one system to another will have his reasons, more or less well informed. But an inexperienced person seeking employment and finding it in some function within a system is in a different situation. If he feels that the job is a fair match for his potentials or if he has no internal drive to exploit his potentials more completely, then he may find the conditions congenial and acceptable. In such cases, he does not feel downgraded.

But if that task calls for robotlike performance and he sees himself as much more than a robot, he will soon feel downgraded. And if he likes the company of serious, competent, intelligent, cultured people, and he finds himself surrounded by fellow workers who, being something less than that, can more readily adjust to existing conditions than he, he will also feel downgraded.

It can scarcely be questioned that ideally, at whatever level, it is desirable for a job to have an upgrading effect. If this is desirable, then it should be an objective of management to provide this to the extent feasible.

It is not suggested that this is always easy. The enormous processing capacity of the computer gives it an insatiable appetite for data, and the economics of the situation usually requires the capacity to be utilized to the maximum. This calls for the preparation of vast quantities of information to be converted to a form adaptable to computer input, and thus a large staff must be employed merely to convert the information from one form into another.

The tasks involved are utterly uncreative; they require, for the most part, only a substitution of one form of symbolization for another. The tasks and routines are easily learned and performed, and the experience gained on such tasks is generally of little value and teaches little about the underlying rationale and functioning of the system. It offers little in the way of preparation for more exacting, more rewarding tasks or for supervisory responsibilities.

Furthermore, the ratio of low-level employees to those in positions requiring greater skill and knowledge is usually great, so that the promotional path would be crowded if many were qualified.

However, the situation in many organizations appears far more downgrading than it need be. There is a regrettable tendency to attract and exploit young new employees, recognizing that the downgrading will be intolerable sooner or later, doing little or nothing to mitigate the effects on the individual, and assuming, expecting, and adjusting administratively to the rapid turnover made inevitable by this practice. Such policies leave much to be desired—on both sides.

Effects on Morale

The imagination-stretching differences in levels of achievement by human individuals have been pondered by philosophers, studied by educators, investigated by scientists, and worried over by managers. It is surely among the most fascinating, challenging, baffling, and essential fields of study; and without some fairly well-founded insights into these complex intangibles the leaders of organizations can suffer serious failures.

We know from the piteous history of domination and cruelty that people can be forced to their utmost degree of physical exertion by the threats to themselves or to their loved ones of death, physical punishment, or the deprivation of the necessities of life. Tyrants have erected some of the world's greatest monuments with the forced labor of their people. Impressive economic gains have been wrung from or maintained by the systematic exploitation of slave labor. The potential productivities of compulsion are more or less understood by, though generally unavailable to, modern industrial cultures.

More familiar to these cultures are those highly motivated persons, ambitious or farsighted, who deliberately deprive themselves of all but the barest necessities and force themselves to the most pitiless exertions in order to achieve some long-range, generally material or tangible, goal, such as wealth or success, for themselves or for their children. Their own initiative thus imposes upon them hardships rivaling those imposed by the most heartless dictators.

On the other hand we have seen explorers, mountain climbers, pioneers, hero-patriots, adventurers, and some scientists subjecting themselves voluntarily—even eagerly—to the gravest of risks and the severest of conditions, generally for the achievement of goals that can bring them only or primarily the most intangible of rewards. Yet their motivation is often maintained at an incredible level.

We find artists and poets traditionally starving in garrets to fulfill their creative destinies; and we know about many others in the widest diversity of fields—including many branches of medicine and the sciences—who have dedicated themselves to achievement in the directions that attract, interest, fascinate, or obsess them—all with great or even total disregard of the materialistic considerations that appear to weigh so heavily with others.

We even find in the governmental and political arenas some apparently selfless individuals motivated by a sense of duty or moral obligation, who sacrifice what could be promising careers elsewhere to pursue courses that they deem—by potent subjective standards—truly worthy of their powers and efforts.

And we observe also that some groups exhibit quarreling, rivalry, malicious gossip, and backbiting, while others show great social coherence and mutual regard. The members of some organizations become cynical and psychologically detached, while in other organizations the members seem united to achieve common goals.

An organization, according to a definition in common currency among businessmen, is "a vehicle for accomplishing the objectives of a group." Such a definition, while useful for many purposes, and having an undoubted, though limited, validity, fails to take into account the widely observable fact that in many, probably most, and possibly all organizations, individuals have their own objectives, quite distinct from those attributed to the group, and that some of the individuals in the organization care little about the objectives of the group.

The traditional prototype of high morale is generally represented in military terms: the indomitable, fearless, dedicated, and highly disciplined men who readily offer their lives to make their personal contribution to the victory of their outfit. The Marines at Mount Suribachi, the confident troops who followed Alexander and Caesar and Napoleon, the sailors of Nelson's fleet, such examples are offered to illustrate the idea of group morale that nourishes success.

A test of morale that is even more impressive than the risk of life to gain victory is the risk or sacrifice of one's life in the face of defeat: "The Old Guard dies, but it never surrenders."

Perhaps no more valid proof of organizational morale can be offered than the willingness of the members, without hope or expectation of tangible reward, to lay down their lives to gain success for or even to maintain the reputation of the organization. But in relatively normal civilian life, such dedication is even approached, if at all, only in the arena of amateur athletics, or perhaps, in a somewhat different way, in some religious orders.

It seems clear that the motivations that drive some executives to overwork and thus to ulcers and coronary attacks are not based solely on the love and devotion they feel for the organization or for an inspiring leader. Clearly, some morale is closely related to conscious self-interest.

What can be learned from all of these matters of common observation, and from the many well-documented case histories, that can be effectively applied to the administration of system operations? What principles, what regularities, what predictabilities can be derived from the experience of others that can help to improve human performance in the implementation of system procedures? What can be done to apply the vastly advantageous leverage of high morale and powerful motivation to the more effective operation of systems?

It seems clear that we must be concerned with at least two significantly distinct aspects of the matter: the behavior of individuals as individuals, and the behavior of the same individuals in groups.

Even the most differing of individuals will have some similar characteristics or behavioral or reactive patterns; and even the most similar of individuals will differ in many significant ways. Thus if we "build the group," we tend to sacrifice potentials for unique individual contribution; and if we emphasize individual contribution, we may undermine group effectiveness. Clearly, an appropriate balance must be sought between teamwork and the "star system"; and this balance must be well adapted to the specifics of the situation. So we must seek to adjust any attitude-affecting measures we can control to the needs and occasions of the individuals in the group, to the group collectively, to the functions to be performed, and to any significant environmental and peripheral factors that we can identify and affect or influence.

The usages of systems ordinarily require that those individuals assigned a common function must bear some degree of collective responsibility for the fulfillment of at least the routine requirements for that function. Thus if there are x individuals employed in a specific phase of a system, and one is sick or on vacation, the normal flow of work must be performed by $x - 1$ individuals. Systems operations, of course, vary in their administrative arrangements, and may be more or less flexible in this regard; but the assignment of a common task to a group, which is a common administrative practice, implies group responsibility for its performance.

If the individuals in the group accept this kind of responsibility, they will tend to accept a normal added burden of work without serious resistance. But if they are oriented to individual quotas of productivity, they will tend to react negatively to any situation that appears to threaten them with a unilaterally imposed increase in the quota. (Piecework or compensated overtime introduces additional considerations that affect individual adjustment or reaction.) And, of course, if the effect of group responsibility for productivity should impose an excessive burden, there will be resentment, and the acceptance of such responsibility will be weakened.

When an abnormal load of additional work is imposed upon a group functioning within a system, the negative reaction will bear some relation to the established character of the normal task. A considerable degree of tolerance may exist for sharing the work load of one absentee; this tolerance may be less equal to the necessity for sharing the work load of two or more absentees. And the occasion for absences will have their special effects: Vacations, scheduled for all, may invoke adequate tolerance, but consistent absenteeism on the part of one or more members of the group can arouse serious resentment. A prolonged absence—however justifiable on medical or other grounds—will often result in resentment at management's failure to provide a replacement.

A particular enemy of group morale is the recurring incidence of deficient input. Material received from another part of the system is, naturally, subject to evaluation as to its ready adaptability to the functioning of the group. If the material is not in suitable form, is difficult to read accurately, contains errors, or otherwise complicates or adds to the burden of the normal work load, there will be resentment. If such conditions persist, this resentment may mount to serious levels of hostility against the individuals or group directly responsible, and against management for permitting such conditions.

It goes without saying that, in most system operations, the conditions considered by the employees in a specific phase of a system to be normal for them may, in fact, never be actually realized. These tend to conform to the standards of the system itself, and allow little for deviations that add to the work load. However, many individuals and groups entertain a double standard; they are far more tolerant of their own conduct and quality of output than they are of others'—and especially of those whose deficiencies may adversely affect their work loads.

It is futile to attempt to compare the situations of groups functioning in systems with the melodramatic pictures of beleaguered, outnumbered outposts of U.S. Marines or French Foreign Legionnaires, maintaining the honored corps tradition of fighting to the last man, with no surrender. The group that processes a phase of an information system has little in common with the desperate defenders of a wagon train attacked by Apaches on the way west.

When life is at stake—when survival is the issue—the disciplined, trained, high-morale group outperforms others because individual emotions and personal impulses are subordinated to consciousness of and confidence in the superiority of coordinated group action. When the stakes are far lower and the factors providing for group cohesion are much weaker, then individual reactions and motivations exert a centrifugal effect. "If the work doesn't get done, so what? No skin off my nose."

It is probably true that a majority of the employees functioning on lower levels of systems operations are under no very potent compulsion to satisfy the requirements of the job. The ready availability of other comparably compensated employment, the anticipations of marriage and pregnancy, the potential recourse to parental subsidy, even the choices of return to educational pursuits or fulfillment of military obligations offer a variety of alternatives that may apply to many individual situations.

Under such circumstances, the job must offer some positive attractions, and there must be some degree of minimization of negatives, if it is not to be merely tolerated on a substandard performance basis. The possibilities for providing such attractions and for minimizing conditions that provoke adverse reactions are limited in kind, and perhaps depend not only upon relevant policies that affect the conditions of the individual/group but also upon the way in which such policies are carried out, and perhaps even more, in some situations, upon the personalities of those who are charged with the implementation of policies. (The roles of leadership are discussed in Chapter 7.)

Unfortunately, systems operations, as generally conducted, offer little to the participants in the way of positives. This has the effect, in connection with high turnover, of tending to staff the systems with individuals entering the work force, those seeking temporary positions, or those with little expectation of intangible rewards for their work. Thus the conditions may prove workable, and are allowed to continue.

Many employees develop a high level of tolerance for such conditions, relaxing to match the minimum requirements for the situation. Their usual rationalization: "All jobs are the same; all bosses are the same; they aren't too hard on you here, and the money is about as good as you are likely to get anywhere else." This outlook may exert a stabilizing effect, but to the extent that it causes employees to remain, it is probably fostering retention of those with lower levels of productivity and minimal potential. The individual and group morale thus engendered is of a low order.

Those employees who feel some need of positives will tend to seek them in personal indulgences, such as stretching coffee breaks and lunch hours, gossiping at the water fountain, and indulging in excessive irrelevant talk while at work. All of this tends to lower the group's standards of conduct and may be highly deleterious to morale.

There is also the diversion of individual interest in building some sort of personal status at the workplace, which is irrelevant to job performance. This often takes the form of peculiarities or extremes of dress and other aspects of personal appearance, and deliberate eccentricity of interpersonal behavior, including various forms of ostentation or aggression, which may be addressed publicly or otherwise to lower levels of supervision.

Other characteristic manifestations of the drive for individuality may even extend to deliberate antiorganizational behavior, demonstrating that the individual "can get away with murder" in terms of low quantity and quality of work, the violation of administrative norms for allowable absence, tardiness, disturbance of others, behavior toward management, and the like. Such behavior may even offer an alternative leadership to the group. (This does not refer to union representation.)

More constructively oriented employees will seek satisfactions off the job, in hobbies, recreational activities, and other interests. A small proportion will apply the considerable margin of energy not absorbed by their work to academic pursuits, to active participation in various external organizational activities, or to a more intensive family life. Unless an appropriate balance obtains, however, such extracurricular activities—however meritorious and inherently worthwhile some may be—will inevitably encroach more and more on the individual's disposable attention, and thus further reduce the proportion granted to the workplace—which presents such ineffective competition.

Nonidentification and Alienation

Group morale correlates strongly with a sense of personal identification with the group on the part of its members. Organizations with high morale are characterized by a relatively high proportion of participants who feel that the organization fulfills or provides for them to fulfill, in some degree, their goals and aspirations. At the same time there must be a substantial degree of compatibility and consistency between organizational and personal value systems; and, indeed, in situations where individuals feel strongly their personal identification with the organization, values tend to be learned or derived from organizational *mores*.

Individuals differ widely, of course, in their kind and degree of need for a sense of identification with the organization that employs them. At one extreme appears the utterly loyal devotee who derives a major part of his own self-regard from his association with the organization. He has only a few minor interests outside of his employment, if any, and can talk of little apart from the shop. His identification with the company governs him so strongly that he feels its progress and its successes almost as if they were his own or in his family. Indeed, his attitude toward the company approaches that of the proud and loyal member of a large and important family. This attitude may even extend to a system in which the individual feels involved.

At the other extreme one finds the individual who merely needs a job and tries to make a reasonably acceptable bargain: his time and effort against pay, benefits, and privileges. As time goes on he is likely to seek to improve his bargain, and he may do this both by seeking greater rewards and by giving less: withholding time and attention from his work. This puts his interests and those of the company at odds; his values are thus opposed to those of his employer; and any existing sense of identification may tend to decrease or disappear unless some more positive element operates more favorably.

It should be clearly recognized that the present value

of an employee to an organization is not related solely to the loyalty and identification of the individual. The most devoted employee may be barely adequate at his job, while the largely alienated employee may perform well when he chooses to do so. Many large office staffs include individuals who demonstrate alienation by a high and unjustified level of absenteeism, but who are kept on because of their relatively excellent productivity when at work. Individual employee potential correlates even less with identification; in fact, the discrepancy between potential and assignment often contributes to alienation.

Employees may be readily alienated by dislike of a superior or resentment of treatment received at the hands of management (these represent leadership effects). Other important factors fostering alienation include environmental conditions—physical or social—which occasion negative, aversive reactions. And another, often primary, factor involves personal reaction to performance of the assigned task.

Such conditions may operate in any employment situation, of course, but their probable relative incidence in systems activities is affected by the relatively high proportion of tasks that offer little or no inherent satisfaction and in which participation appears to confer little or no associated advantage or value of learning or experience.

The individual who continues to associate himself as an employee with an organization in which he feels alien and who remains in employment that offers little or no satisfaction subjects himself to daily exposure to experience—occupying necessarily a major proportion of his working time, energy, and attention—that must require continuing unilateral adjustment on his part.

Such adjustment may take many forms, and, of course, some individuals will require adjustment more than others. There are always, at any level, those employees who seem to adapt themselves almost perfectly to the job, seemingly having no aspirations beyond the present situation. In such cases, it may be that the individual recognizes that the requirements match his potentials.

But, on the other hand, if the individual lacks drive, he may merely relax and be unwilling to make efforts to engage his potentials more fully than the present occupation requires. Such cases are not protected against frustration. If the individual comes to realize not only that his potentials are wasted but also that he has not done all he could to improve his situation, he will come to feel a kind of guilt, a self-reproach that can lead both to self-contempt that impairs his effectiveness and to resentment of the organization that has allowed this to occur. Either or both will increase his coefficient of interactive friction to the detriment of organizational morale.

The most common form of adjustment involves the belief that the available alternatives are not, or not substantially, better than the present employment. Such evaluations can partake of the complexity suggested by the exponential effect of many different considerations: levels of compensation; opportunity; status; convenience; hours; personal tasks, preferences, and inclinations; evaluations of and feeling for one's superior; social considerations and myriad others. Each of these has different weights and combines in different ways for each individual, changing for each individual with time.

Such an adjustment, of course, rests substantially on arguments that appear to reflect negatively on the individual's employment. To say that "A change wouldn't make any real difference" is a far cry from saying "I like it here."

Furthermore, the balancing of alternatives is sometimes weighted by tangible considerations. A youth a few months out of high school or college, in his first job, may have little to lose by quitting. But an employee with some years of seniority is in a different position, and if he has a substantial interest in a pension that would have to be wholly or largely sacrificed, he may well feel that he cannot afford to quit, however dissatisfied he may be with his job.

Individuals in such situations often describe themselves as trapped or in dead-end jobs and may become extremely embittered. In some cases they compensate by becoming anticompany propagandists, constantly grumbling and complaining, and seeking to initiate and exacerbate the disillusionment of newer, younger employees. Their own alienation thus spills over into a threat to the morale of others.

This condition represents, of course, a serious failure of adjustment. It should probably be considered as a failure that began, unnoticed, years before. But such an employee's sorely tried supervisor may find it little consolation to realize that years ago some other member of management somehow helped to bring about his present problems.

Perhaps few supervisors in such a position will dwell long upon the realization that the alienated employee is unhappy, that his unhappiness is chronic—a tragedy—and that the organization must bear its share, whatever that is, of the responsibility and the blame. Other employees may dislike and resent the disgruntled one, but they will dislike even more seeing an older, somehow pitiful fellow employee dealt with in a manner they feel to be arbitrary, unfeeling, unfair.

It must be assumed that, when a preferable alternative is offered—a better job—and there is no preclusive sacrifice involved, the individual will naturally move toward the more attractive opportunity. (The frustration of the trapped employee is, at least in part, due to his lesser freedom to change jobs.) A high rate of turnover may be due, in greater or lesser part, to such movements; but if so, the relative unattractiveness of the situation departed from should be apparent.

Some large organizations, as a matter of policy, hire almost any available applicants at the lowest feasible levels of compensation. Such a policy requires being set up to operate with relatively inefficient, low-cost labor, and they accept a high rate of turnover as an inevitable, planned-for element in their way of doing business. This kind of situation is often found in offices, retail establishments, and manufacturing operations where low levels of skill and experience can be utilized routinely.

Tasks are designed with unexacting requirements, and supervisors bear the burden of eliciting at least the minimum planned contribution of productivity from the uninterested, often grudging, employees. In conception, such situations deliberately ignore the potentials of the individual employees, treating them with impersonality and evidencing only expectations that can readily be met by all but a few. Managements conducting such operations are like farmers or miners who set their goals at levels far lower than those attainable and are satisfied with their planned yields, wastefully leaving much of value behind in field or mine.

If we consider these positive opposites of alienation and nonidentification—involvement and personal identification with the organization and its objectives—we touch upon the other extreme of the spectrum which covers the range of a major source or reinforcement of constructive motivation.

Let us recognize that personal considerations will always remain the paramount motivational focus of any individual. Consequently, it seems obviously to the benefit of organizations to enlist the force of personally oriented motivations, thus attracting a greater share of the individual's interest, energy, and attention; inducing the devotion of a larger proportion of his constructive effort; and developing and utilizing an increasing share of his applicable potential. To accomplish this, the organization must provide opportunities that enable individuals to achieve their personal goals (or approximations or equivalents thereof), and also create conditions that make it possible for aspiring individuals to achieve, to a satisfactory degree, a sense of self-fulfillment.

To the extent that such conditions exist, the organization may find the relevant resources of individual employees to be, in effect, its own, willingly and voluntarily and adaptively applied to the business of the organization at the initiatives of the individuals themselves.

To the extent that such conditions do not exist, the organization will suffer the loss of those who would most value such conditions (generally those employees with the greater sense of personal potential), and the alienation and nonidentification, at least in some degree, of those who remain.

The difference will amount to far more than management can envisage (and, perhaps, afford).

The substantially alienated individual may seek recourse in withdrawal behavior or in hostility.

Constructive withdrawal involves quitting to seek more acceptable work. But there are other forms of withdrawal, such as the withdrawal of all but a minimum effort to meet job requirements. A more thorough withdrawal can be a withdrawal, perhaps, from life. This too often takes the form of destructive self-indulgence, sometimes resulting in alcoholism or even drug addiction.

Hostility, as a manifestation of alienation, may take the form of generalized, idealized antipathy, as in antiestablishment roles, interests, and activities. Sometimes this takes overt forms such as in obvious hippiedom, proselytizing for various antiestablishment causes, extensive participation in protest activities, and other forms of militancy.

But alienation leading to hostility may also result in the highly practical form of participation or even leadership in union activities. A hostile shop steward, who seems to enjoy creating difficulties, cannot be explained solely by his love for the ideals represented by the organization of labor or by his interest in winning higher wages and better working conditions.

Unfortunately, many of the conditions current in systems operations tend to foster alienation in all but the most unaspiring and resigned. They are often havens for the underprivileged—blacks, Puerto Ricans, recent immigrants or their children—and for the underachievers, although many of them may have a sense of achievement and even of status when they compare their situations with those of parents, siblings, or neighbors. For those with some upward drive, to feel oneself progressing upward tends to build morale, despite some adverse conditions. But the depersonalizing, alienating conditions also process large numbers of the young, who are just entering the work force, and to alienate them is to render a distinct disservice to the community.

A Certain View of Systems

Any one system operation, or system operations generally, can be perceived from varied points of view and can be evaluated according to a wide variety of standards and values. Unfortunately, the standards and values of those who authorize, design, install, and control system operations, and especially of those who call them into being, tend to differ vastly from the standards and values of many of those who find employment in the operation.

The system is created, we must assume, to provide a more advantageous means for accomplishing certain objectives, primary or contributory to the basic functioning of the organization in pursuit of its fundamental goals. The ancient principle of division of labor, properly coordinated through modern techniques, is supplemented by advanced and highly productive equipment or—increasingly—utilization of advanced and highly productive equipment that calls for ancillary activities, usually of an extremely limited, prescribed, routinized nature, also involving the division of labor—sometimes to an extreme degree.

Thus one thoroughly legitimate and creative, even essential, point of view is represented by the individuals who first recognize the need or perceive the opportunity for a system, who identify and evaluate the potential advantages, and who perform the planning, organizing, and executing necessary to bring it into being. But this point of view too often overlooks many other significant aspects of the matter, such as the effects upon those employees, supervisors, and managers more or less directly involved in the operation of the system, and upon the organization as a whole.

We know that, unfortunately, many system installations have proven unsuccessful, and some even calamitous. We are fully aware that almost all system installations require substantial revision and adjustment, in situ, before they can begin to function effectively for the purposes and in the manner intended. All this suggests that those who conceive of, plan, design, and install systems are far from infallible, and may even be expected to overlook significant factors that must affect results in unforeseen ways or even produce their own effects. And we are justified in assuming that such results or effects are most likely to be unforeseen or disregarded in direct proportion to the degree to which their relationship to the specific and directly related functions of system operation is incidental, peripheral, or remote.

On paper, an "operation" represents a specifically defined activity, accomplished in a prescribed way within definite limits of time and cost. On paper, a "job description" represents a function or a set of activities that someone anticipates will be performed as envisaged. On paper, complications, side effects, and other variations, developments, and tendencies do not appear, and are irrelevant, however inevitable. And systems are planned on paper.

In the useful and highly applicable terminology of Alfred Korzybski, founder of General Semantics, plans for the operation of systems are *maps,* and they should not be mistaken for the actual *territory*—in this case, what really happens.[4]

[4] *Science and Sanity,* 4th ed. (Lakeville, Conn.: International Non-Aristotelian Library Publishing Company, 1958).

Many systems, of course, operate successfully. They contribute invaluable efficiencies and economies, and they provide hitherto unavailable but now indispensable services. The already realized advantages and benefits of systems are enthusiastically acknowledged, overbalancing the failures and the disappointments. And their potentials for even greater contributions are eagerly explored.

The two basic questions raised here are: How can systems be best served by the indispensable human participation? And how can systems be made to operate with the least negative effect (and, of course, if possible, the greatest positive effect) upon the human elements of the organization directly or indirectly involved in the systems? Of course, the two questions are related, and any workable answers must be functionally related. Unfortunately, while some attention is necessarily paid to the first question in all system situations, the second one is largely disregarded, and the result of this disregard must inevitably impair the effectiveness of whatever answers are provided to the first question.

If an attempt were to be made to state basic objectives as guidelines for the development of answers to these two questions, they might read as follows:

1. Individuals assigned to tasks in system operations should meet standards of efficiency and accuracy.
2. They should accept irregularities in the work load, however caused.
3. Vacancies in supervisory and managerial posts within the system should be filled, so far as feasible, by promotion of competent candidates from within the system.
4. Personal potentials of individual employees should

be recognized, fully utilized and developed, and properly applied to the requirements of the system and/or of the organization as a whole.

While the principles suggested in these theoretical objectives have long been recognized as applicable to organizations generally, it is a matter of observation and judgment that not nearly enough has actually been done to realize them, and perhaps least of all in the administration of many system operations.

Too many administrators have set their goals so low as to attempt merely to keep the operation going with a constant flow of new personnel, junior and unsophisticated, but low priced, while accepting the inevitably concomitant disabilities of low productivity and high turnover. Some large clerical operations are much like certain marine animals that live by ingesting and expelling great volumes of seawater, nourishing themselves by straining out a few small organisms with each cycle.

The psychic and psychological cost to many of the individuals, and the cost to some of them in wasted time, is not calculated in such policies, and the loss to the organization in unused human potential is apparently not even considered in many such cases. Perhaps, if serious, constructive thought were directed into these areas, some of the policies now governing some system operations would be seen to be so costly in their overall effects as to reveal that the systems appear to run at an intolerable overall loss.

The system staffing should be appropriate to the tasks assigned on an individual basis.

Systems personnel should be motivated to perform their assigned tasks with a high degree of efficiency.

Since quality (or accuracy) is a primary requisite of most systems, personnel should be concerned with these and be capable of meeting and be motivated to meet system standards.

Systems personnel should take an interest in their work, in the relation of their work to that of others, and in the successful operation of the system as a whole. They should be appropriately cooperative, and understanding of and accommodating to irregularities in the work load.

An adequate proportion of system personnel should be capable of and interested in moving up into supervisory and managerial positions.

7 The Individual and the Task

BASICALLY, the typical office is suffering from inadequate adaptation to change. Great increases in both the number and variety of transactions have added to the volume and complexity of operations. The introduction of mechanical, electrical, and electronic equipment has changed the nature and the requirements of much of the work. In spite of the contributions of technology, the size of the work force has had to be greatly increased. And socioeconomic changes have greatly altered the character, motivations, and cost of the enlarged work force—now largely recruited from young and inexperienced personnel, many from non-white-collar backgrounds.

Under these conditions, the ever smaller proportion of more capable employees must carry the burden of most nonroutine activities, and this includes the responsibility for correction of much of the ever-increasing proportion and volume of error. At the same time, complexities of operation make many errors far more difficult to correct.

The growing burden of urgent nonroutine work so fully occupies the time of those capable of handling any part of it that other matters tend to be neglected, and the worst impact of this condition is felt in the deficient training, informing, and handling of personnel, especially those newly inducted and recruited, who are relied on for a great volume of what should be routine.

The implementation of basic policies of personnel administration and of effective supervision is neglected or inadequately conducted by managers and supervisors who may themselves be insufficiently trained for such functions, but who are also too strenuously pressed by their own personal work loads to apply what they might otherwise be capable of bringing to bear on the situation.

This fundamental deficiency results in maladministered inductions, inadequate training, neglectful supervision, and other conditions that adversely affect the initial and subsequent behavioral patterns of individual employees. The result is seen both quantitatively and qualitatively in task performance.

This deteriorating condition in turn requires both more and better supervision, and the addition of personnel capable of catching and correcting errors, neither of which is forthcoming. The resultant disaffection leads again to inferior employee performance, well below most individual capabilities, and thus the degenerative cycle continues.

Management's Role

Such situations call for arresting the cycle of deterioration. More and better supervision, better adapted training and induction procedures, and greatly improved administration of personnel policies are all necessary. The results can be substantial as the potentials for improved employee performance are truly great.

It is unlikely that many managers can envision adequately the gratifying improvements that are obtainable. A quantum jump in productivity and in the error-free quality of production can be achieved if a greater proportion of employee capability can be focused on the required tasks.

To attain the indicated potentials is a real challenge to management, but the way is indicated; and failure to take appropriate measures will involve increasing penalties. Increased dependence upon automation, such as electronic data processing, can provide only part of a solution: The inputs and outputs must still be handled by employees whose performance is subject to influences that can drastically affect results.

Organizations need to utilize more, not less, of the potentials of their people on every level. This urgently requires action toward bringing about conditions that will foster performance on a higher level of error-free productivity.

While the discussion has been directed primarily at clerical situations, the principles are generally applicable to manufacturing and service operations where individual contributions have been narrowed and routinized in such a way as to require, supposedly, a minimum of knowledge, skill, or application of individual ability.

The behavior of any individual at any given moment may usefully be regarded as a *gestalt*—a whole or total event not fully or adequately represented by the sum of its identifiable parts. On consideration, it appears to partake of the nature of happenings, contributed to by many dynamics, perhaps largely unknown; and affected by many factors, some static and some dynamic.

Thus the personality of the individual, acting in response to inner needs and drives and in reaction to outer stimuli and conditions, is affected or guided by external factors that may be relatively static (physical elements in the environment, or fixed rules, directives, and instructions) or dynamic (other personalities, active supervision, new policies, or changing environmental conditions).

The individual whose work is wholly supportive of and subordinate to a closely organized system finds himself in a situation where his desired activities for useful output are rigidly directed and controlled. At the same time he may feel the impact of other potent influences of a highly dynamic nature, however irrelevant to his assigned function, arising from internal drives and frustrated reactions, from the manifestations of supervisors and fellow workers, and from emergent events and hour-to-hour experience.

It would be rare, if not impossible, for any one known influence to account fully for any significant behavioral development, although a more completely causative effect may be attributed to a proximately causal event that should then more properly be regarded as triggering.

It seems appropriate, in the attempt to evolve or elicit useful principles and guidance, to consider the factors that may influence the behavior of employees operating within organized systems under several general categories:

1. Management Effects
   a. *Dynamic factors*—supervision, training, and planning.
   b. *Regulatory effects*—organization structure, work

flows, standards, procedures, job/position descriptions, work loads, and methods.

c. *Personnel policies and practices*—recruitment, selection, assignment, placement, transfer, career opportunities, career planning, job rotation, compensation, incentives, appraisal, and promotion.

d. *Human engineering factors (biomechanics)*—equipment, space, physical arrangements, light levels, noise levels and effects, crowding, distractions, materials, processes, and man–machine relationships.

2. Joint Effects: leadership, pressure/pace, pace/pressure, social environment, mores, group morale, ambiance, and tradition.
3. Individual Effects
   a. *Contribution*—capability/requirements ratio, capability/performance ratio, quality of output, and social factors.
   b. *Subjective factors*—attitudes, individual orientation, individual morale, motivations, identification, conditioning, self-image, and role perception.

While the above outline provides potentially useful (tentative) categories of "factors" affecting behavior, these factors will be discussed under a somewhat different organization of subject matter, hopefully more readily applicable to specific and practical problems.

These factors will now be discussed in terms of their applicability to specific problems. Grouping the factors in this manner may be useful to those facing actual problems in optimizing the relationship between individuals and systems.

Adjustment of Tasks to Individuals

In accordance with the approach suggested above, the adjustment of tasks to individuals in system operations will be considered under two disparate headings of generally parallel significance.

The first heading has to do with "Regulatory Factors and Effects," and deals with all manner of communications emanating from management which are intended to, or do, specify, modify, alter, or otherwise affect a task or the manner in which or means by which or conditions under which a task is to be performed. This would also include conditions of compensation, standards of quantity and quality of production, safety requirements, and so on.

The second heading has to do with "Human Engineering Factors," and deals with the effects upon task performance of physical elements and conditions—planned or unplanned—which do or may have significant effects upon task performance.

Regulatory Factors and Effects

Management, in establishing an organization, initiates plans, policies, programs, directives, instructions, job/position descriptions, procedures, methods, and other formal or informal communications intended to order the functioning of individuals and groups. If the organization is regarded as, in some sense, a system with subsystems, then these communications constitute, represent, or symbolize the effects that convert the discrete elements into a whole that can reasonably be considered a system.

Such communications are used also to initiate desired changes in organizational and individual functioning. But once functioning has been adjusted to the initiative, their effect is relatively static and would normally operate toward maintaining the status quo that, after their execution, they represent. Thus the fixed item of management initiative becomes essentially regulatory in effect and tends to serve in continuity as a preserved standard for the functioning to which it may apply. Gradually, as other conditions and factors change, it becomes obsolete.

Thus management initiatives (with the exception of those aimed at limited occasions or with specific expirations—and even these may create lasting precedents) tend to follow a pattern of life-cycle effectiveness beginning with the dynamic agency of change, relapsing to a regulatory function, and then gradually becoming obsolete, that is, inapplicable or, at least, less beneficially applicable, or less acceptable to some or all of those affected.

Such obsolescence, inapplicability, or unacceptability is seldom adequately anticipated and often goes unrecognized, or at least unacknowledged by management—or unacted upon—for extended periods of time. Even when recognized and acknowledged, there is often considerable (and, to subordinates, inexplicable) delay in official action to correct the situation.

During any such delay, of course, the regulatory effect will generally have increasingly negative value, sometimes approaching the destructive in its effect. In too many instances, necessary changes are not initiated until seriously disruptive experience has forced management to raise the priority previously assigned to consideration of the matter. Thus do some systems deteriorate, or at least allow damaging internal maladjustments to occur and develop.

Examples of such incompatibility between regulatory effect and situational requirements range from the traditional Code of Military Justice, finally recognized as unsuited to a peacetime draftee army at a time of widespread antimilitarism, to the practice of having municipal bus drivers (on one-man buses in some cities) carry cash and make change for passengers until too frequent holdups finally dictated the locked coin collection box and no-change policy.

Every organization rests upon and operates on the basis of many, many formulations (written or otherwise) that channel (more or less effectively and specifically) the activities of its members. These constitute or embody the patterns that should guide systematic functioning. But in the very nature of the situation, it is impossible for all of these formulations to be kept constantly at the peak of useful applicability.

Extreme examples are to be found in the legislation governing our nation, states, and cities. When New York City's basic charter was revised in the 1930s, it was found to embody some ordinances dating back to the eighteenth century, some even quoting fees and fines in terms of shillings. Many dealt with such matters as the care of patrons' horses at inns, the number of persons permissible per bed, the operation of public horse cars, and the like. Others embodied more damaging obsolete governances of police practices, jail administration, tax collection, fire protection, and education.

Many of our leading corporations, states, and municipalities still have in circulation organization manuals with comparably, almost ludicrously, obsolete provisions, and with less excuse.

Regulatory effects, when at their best, serve to guide individual efforts along lines desired by management. When up-to-date, well designed, and appropriately prepared for and appropriately introduced into the organizational structure, they contribute to individual productivity and improve organizational functioning, but they also have valuable intangible effects. They build the individual's respect for and confidence in management, and thus contribute to attitudes of positive value that foster identification and generate and support morale.

As such regulatory communications remain static and unchanged, other conditions are changing, and the considerations that brought them into being in their original form become less and less compelling as other considerations arise to which they may not apply. The positive regulatory effect, which may have had substantial value, deteriorates. Gradually or suddenly, it may even become negative in effect, and thus in value.

As the effect deteriorates in terms of usefulness, applicability, and acceptability, the regulation actually invites impairment of system functioning. A common example is the promulgation of a carefully devised procedure for an operation. A change in the equipment used, in the input (material or information), or in the output desired will require a change in the procedure. If the procedure is not suitably modified at the time of change and by the same level of management responsible for the initial procedure (or at least at a level higher than that directly responsible for the operation), then some degree of improvisation will be necessitated.

This development loosens the authority of the still applicable elements of the initial procedure and tends to initiate a trend toward departure from the originally conceived and designed pattern of the operation. This may occasion deterioration in the operation, which may proceed for a long time. If the locally improvised procedure actually improves the operation, that fact will tend to inspire a generally negative attitude toward all procedures subsequently handed down from above. Another kind of result is illustrated by the proliferation of printed forms of all kinds that often clutter the clerical processes as they originate in the lower echelons of systems to meet locally felt needs.

A common and deplorable manifestation of the faulty relationship between regulatory structure and efficient, implicitly approved functioning is seen occasionally in the slowdown—a form of labor protest in which the employees work by the book, following each regulation literally. When this results in a breakdown, a near breakdown, or a severe deterioration of operating efficiency, it is clearly seen that the map of regulatory verbalization imposed by management does not correspond to the territory of desired and approved or even necessary functional behavior of the employees.

Another example is found in the common experience of those attempting to administer safety programs. Some of the safety regulations imposed are seen as unacceptably onerous—they require extra activity or make an activity more difficult or tiresome. Such a regulation may soon be observed only in the breach, but the disregard of one regulation that is part of a system leads to disregard of other parts of the same system, and thus a whole program sometimes lapses into ineffectiveness because of the unacceptability of or failure to enforce observance of one element. Such a development casts a shadow over subsequent efforts by management to affect or alter behavior patterns, making more difficult the process of initiating desired change.

Thus a system involving people, once constituted, is

maintained by patterns of individual behavior generally prescribed, specified, and invoked by regulatory measures, and the maintenance of such measures in some appropriate relationship to the realities of operations has an importance that parallels in value the importance to the organization of the system itself.

Human Engineering Factors

The performance of individuals, in or out of systems, is inevitably affected directly or indirectly by physical factors in the immediate or general environment. These factors may be as direct and apparent as the efficiency and controllability of equipment, or as indirect and intangible as the decor of the workplace.

The operation of equipment is more or less directly affected by at least sixteen classes of factors:

Inherent efficiency of equipment
Adaptation of equipment to specific needs
Availability of equipment
Accessibility of equipment
Maintenance requirements
Adaptability of equipment to workplace
Applicability of equipment to specific tasks
Operational requirements
Skill requirements
Authorization requirements
Environmental effects of equipment operation
Psychic effects of equipment
Directly related conditions of use
Indirectly related conditions of use
General environmental factors and conditions
Biomechanics

While many of these items are normally comprehended under sound engineering practices, the significance of such factors in terms of their effects on individuals as participants in systems merits special comment.

Equipment multiplies human effectiveness, and thus provides a basis for an enhanced sense of accomplishment. It also furnishes the obvious key to increased productivity, which at once makes it economically feasible and provides the justification for increased operator earnings. A system with little or minimum equipment tends to be a low-grade system, one in which participation brings little status except in unusual cases where there is a requirement for a high degree of individual knowledge and/or skill. A system built around highly efficient equipment tends to shed an aura of prestige upon its human coparticipants. Any impairment of the status of the equipment tends to be reflected in the equipment-derived sense of status as felt by the individuals involved, and often by others.

When the newest navy ship is commissioned, members of the crew are likely to feel a degree of pride specifically associated with its unique status, conferred by its ultramodern design and equipment. As time goes on, they can derive less and less satisfaction from their association with this vessel for such reasons. And as the ship approaches obsolescence, the morale of the crew must be bolstered against adverse comparisons with the newer ships.

Even so, a clerical operation built around a new and distinctive computer—or even desks newly provided with computer terminals—will have a psychological impact upon the employees in terms of perception of self and often in terms of perception by others. As the years go by, the effect fades, and the nearby appearance of a computer of a later generation tends to signal the effect's demise. But the individuals associated with the older computer are still conscious of its continuing multiplier effect upon their productivity.

Because systems all but invariably involve men–machine interaction, and since such systems are almost always designed around the equipment that makes the system possible, the optimization of the man–machine relationships is desirable, not only for the maximization of production efficiency, but also for the generation of psychological effects of positive value. The listing of sixteen classes of factors should be seen in this light.

Inherent efficiency of equipment. Employees are always—though more or less acutely—aware of the inherent and especially the comparative efficiency of equipment. A new employee may be merely conscious that the equipment is old, but a longer-term employee will usually be generally aware of the existence of more modern equipment and may know of more up-to-date installations elsewhere within the organization or outside it. The installation of a shiny new piece of equipment is almost always a source of satisfaction to its prospective users.

An assigned-use machine, as a typewriter for a particular typist, has a rather specific effect upon the intangibles of the individual's sense of his relationship with management. A machine for use by a group, that is, a copier, an adding machine, or calculator, is often regarded—perhaps explicitly—as an indicator of the status of the group or of its function.

Thus the sense of participation in a system can be rendered more or less acceptable by adjusting the degree of appropriateness of the equipment provided. An addition or substitution of equipment that improves individual or group effectiveness for system purposes will have a positive effect upon attitudes, while a belief that the equipment provided unnecessarily limits effectiveness will exert adverse effects.

Adaptation of equipment to specific needs. The individual or group will be sensitive to the implications for the importance of their functions within systems of the equipment provided. The provision of general purpose equipment does not imply system involvement so strongly as more specialized equipment, and equipment that is actually system linked is especially significant in this respect. The presence or absence of a telephone on a desk tells something about the individual's relationship to one or more systems and may constitute an indication of status. The individual's assignment to an input or output terminal of a computerized system provides a clear indication of the individual's function within the system, but also imposes upon the individual a heightened consciousness both of system involvement and of specificity of function. The effect will be stronger than that engendered by the same degree of system involvement and specificity of function without actual physical linkage.

Thus a new employee of an organization may be trained as a typist and assigned to type out certain data on sheets of paper that are transmitted to other locations for processing. At a later stage of system development, she may type the data on a machine that cuts tape for transmission to the other locations. And at a still later stage she may be typing the data on a machine that is cable-connected to a computer. In all three cases she is physically performing virtually identical tasks with parallel functions, but the subjective effects of more intimate, more tangible involvement in the system will be different.

Different individuals will react differently and in different degrees to such changes as these, but the adverse sense of loss of autonomy, even of individuality, can be countered on that level only by a sense of the significance—even the importance—of the functions performed.

On managerial levels, the ever more complete linkage of operations to higher levels of management via data processing—especially in real time—often detracts from the lower-level manager's sense of authority and responsibility, even without specified changes in these. The field manager, for instance, often suffers a sense of loss of independence, which may impair initiative. The obvious remedy is to convince him of the potential benefits and advantages to him from the greater availability of information and of some added degree of guidance. Similar considerations may apply to salesmen and other individuals functioning under limiting controls.

Availability and accessibility of equipment. Queueing effects in the use of equipment generally involve psychological results. A clerical employee is asked to make a number of copies. She proceeds to the departmental copier and joins the line of those awaiting their turns for its use. The equipment thus becomes a social center, competing successfully with the water cooler and the washrooms for this distinction, with the advantage of being functionally related to task. After a few such unproductive but recreational sojourns, the employee's attitude toward socializing on the job may be substantially affected, and any residue of discipline relating to coffee breaks may be seriously impaired. Some supervisors recognize this to such an extent that they sometimes award assignments involving visits to such facilities accordingly.

Inefficiencies caused by deficient availability or accessibility of equipment are often seen as oversights by management or as evidence of the low value placed by management upon employee productivity. The effects can be extremely adverse. Similar observations apply when obstacles are met and delays are incurred in seeking essential supplies.

Maintenance requirements. When failures of maintenance of equipment result in unavailability of the equipment that directly affects employee productivity, employees will see this as a clear failure of management and also as a derogation of the importance of their functions. When employees functioning within a system are to any degree dependent upon the functioning of equipment, there is a tendency to compare the degree of management attention devoted to both and to draw inferences. While a pattern of great attention paid to equipment and little to employees will be resented, the neglect or deficient functioning of equipment will also be resented, though for somewhat different reasons.

An elaborate and expensive piece of equipment may deliver an impaired output due to a minor failure of upkeep; for instance, a printer terminal may lack a new inked ribbon or the type may not be clean. Employees may then resent being thus thoughtlessly, unnecessarily handicapped.

Adaptability of equipment to workplace. It would seem a truism that equipment should be placed conveniently for those who are to use it so that they may work without inconveniencing others. Yet the pressures for space and the problems of space allocation often result in location of equipment that violates this obvious precept. Employees resent, and have their efficiency impaired by, proximity to equipment that they use seldom or not at all, especially when such equipment generates traffic or otherwise offends or distracts. Such conditions suggest to employees both poor planning and lack of consideration for them. The first impairs respect for management, the second inspires resentment.

Applicability of equipment to specific tasks. It seems clear that employees will react negatively to being required to perform specific tasks if the equipment provided for the purpose is inefficient, produces poor results, causes extra trouble, or otherwise reflects unfavorably on management's judgment or on its consideration for them. This kind of reaction will occur far short of the "making-bricks-without-straw" stage.

Operational requirements. The operation of some equipment imposes physical requirements that may be less than acceptable. The need to stand for extended periods is generally resented. The need to reach, to work in a bent-over position, or other inconveniences of operation engender dislike that may build up to important dimensions. The need to exert effort repeatedly or the need to press hard repeatedly with a finger can arouse serious aversion in some employees. Arrangements that strain the vision of the operator are resented.

All such conditions, inducing negative attitudes, exert adverse effects upon employees' participation in system operations.

Skill and authorization requirements. The operation of much equipment requires a degree of skill usually derived from familiarity with the equipment and experience in its use. The acquisition of such capability is a legitimate objective for an employee who works in some visible relation to the applications of the equipment. Failure to provide such training can be taken as a deliberate discrimination against the individual, whereas the proffering of such training at the earliest appropriate moment should have a morale-building effect.

Similarly, when specific authorization is required by management for the operation of equipment, the door is opened to the likelihood of adverse reactions to the withholding of authorization from some individuals. This will be true even when, and perhaps especially because, a certain degree of responsibility is considered a prerequisite and when access to a machine must be rationed or limited for other reasons. Factors affecting compensation for added duties or skills may complicate such situations.

Environmental effects of equipment operation. Some equipment is quite noisy. Other equipment may vibrate, or give off odors, or radiate heat, or otherwise affect the environment. Equipment effects are often distracting, sometimes unnecessarily so. As an example, telephones on the desks of supervisors or senior clerks may greatly and unnecessarily impair their efficiency by putting them in the position of serving as the channel for messages or data intended for others. The loud and constant ringing of a telephone can be extremely annoying, even to those at nearby desks. And the frequent calling of individuals to answer the telephone can be quite distracting to others. Such conditions engender a sense of confusion, which is the antithesis to the impression appropriate to a well-conducted system operation.

Employees vary greatly in their sensitivity to environmental factors, but when their reactions are adverse, there can be a cumulative effect, with decreasing levels of tolerance. Adverse reactions to the environment of the workplace tend to grow into generally adverse attitudes toward the system, with impairment of performance within the system as a likely result.

Psychic effects of equipment. There can be satisfaction in the use of well-functioning equipment, and this satisfaction is acutely felt by some, while others may be relatively oblivious. There can also be satisfaction—of varying degrees of appeal—in working with expensive, elaborate installations: they seem to reflect upon the importance and power of the individuals who operate them. And there is satisfaction of a kind in knowing that the product of one's efforts is more significant because it is greatly multiplied through the application of highly effective equipment. All such effects are attainable in systems, along with their opposites.

Directly related conditions of use. The complex of factors and the interrelations of factors immediately associated with equipment operation should not be ignored. For instance, if an employee who wishes to use an adding machine must first go to another area, and then wait in line, the effect is more likely to be a multiplication than an addition of frustrations. If he then runs out of paper tape, the effect is likely to be far more than merely cumulative, and if he is further involved in delays for requisitioning or procuring the necessary tape, something like an exponential effect must be anticipated. On the other hand, when an employee finds a machine properly maintained and cared for, dusted and supplied, ready to go, he may well derive a sense of system support and of excellent coordination of competent activities that will exert a positive effect upon his attitudinal state.

Indirectly related conditions of use. There are usually factors that affect attitudes that bear only indirectly upon the use of equipment. An example would be the availability of a new machine for use in a situation where a number of similar machines are assigned to individual operators, as in a typing pool. Who will get the new machine? Will it go to the senior employee? Will it go to the employee now using the oldest machine? Will there be a more general redistribution of machines? Whatever course is followed, individual reactions will result that will have some bearing on the individual's subsequent attitudes toward and functioning within the system.

General environmental factors and conditions. Factors and conditions in the general environment exercise their effects on behavior and task performance of many kinds, with or without the involvement of equipment. These effects range from such basic considerations as light level, temperature, ventilation, noise level, and other physiologically significant conditions to such amenities as decor and color scheme, carpeting, growing plants, style and condition of furniture, and shape and size of space. Crowding is intolerable to some, readily tolerated by others. The absence of windows bothers some people but is less important to others. Other conditions are similarly varying in their effects.

There is often great variance in preferences for controllable environmental conditions, such as room temperature, air-conditioning setting, or the opening or closing of windows. Because many environmental conditions must in the nature of things be shared, it is inevitable that a given set of conditions will satisfy some and be unsatisfactory to others. The adjustment of such matters represents both a challenge and an opportunity to supervisors.

Some environments offer a variety of conditions through compartmentation or through variations in the effects of various factors in different locations. Thus placement nearer to or farther from radiators, ducts, fans, doors, and the like may alleviate dissatisfaction and demonstrate consideration for the individual, both helpful in relieving the essential impersonality of system operation.

Factors and conditions involved in equipment or in the physical environment of the workplace are generally under the control of management, and employees can usually do little to bring about more satisfactory adjustments. For this reason some employees are sensitive to these matters in a way that differs basically from sensitivity to task or to the impact of supervision.

By knowledgeable and constructive attention to such considerations, management can create elements of an image favorable to the development of employee attitudes conducive to more satisfactory adjustment to participation in system operations.

Biomechanics. This subject has to do with the design of equipment and its adaptation to physiological considerations of usage. Minimization of fatigue is a primary factor as is facilitation of essential sensory interrelations (readability of dials and other indicators, the adaptability of effectors—knobs, buttons, levers, and other controls—to convenient and careful movement). Improved efficiency and accuracy can result from successful applications of biomechanics to the workplace, and so can improved employee acceptance of working conditions.

Adjustment of Individuals to Tasks

In systems operations there is usually little or no difficulty in determining and specifying functional requirements at each stage of operations. Once the procedures have been formulated and the methods for following them developed, the required performance becomes apparent, and thus the necessary capabilities are readily known. Capabilities, however, often would best be considered and evaluated under at least two headings—operational and temperamental.

At lower levels of systems operations the required operational capabilities are such that they can usually be determined—at least as to meeting minimal requirements—by objective tests or by evaluation of the individual's training and experience. Questionnaires, interviews, and the checking of references generally provide information adequate to determine whether or not the individual offers a probably adequate capability.

It should be more commonly recognized that these investigative procedures generally do not indicate the existence or extent of any capabilities beyond those required for the intended function, nor do they indicate the characterological, cultural, and temperamental adaptability of the individual to the performance of the function and to the total environment in which he will be required to work.

Individual Contributions

As an inevitable result, it is quite possible for the responsible functionaries in an organization to recruit an individual, assign him to a specific task, and train him to perform it, despite the ascertainable fact that the individual possesses capabilities well beyond those required for the task, and is temperamentally quite unsuited to the assignment.

Thus the organization suffers the dual penalty of excessive turnover and of losing a potentially valuable employee; and the individual suffers frustration, disgruntlement, a sense of wasting time, and an aversion to systems operations—if not a growing disillusionment with business organizations. How frequently this occurs will vary, of course, with the organization, the system, and the other major variables. But there is little doubt the occurrence is regrettably frequent.

Capability–Requirements Ratio

One way to alleviate this undesirable situation is to introduce the concept of a capability–requirements ratio into the recruit evaluation process, to invest a greater degree of investigatory effort into making far more real the ratio attributed to each individual, and then to utilize the resulting evaluation in a way that is both useful to the organization and fair to the individual.

A policy and practice of this kind is recommended in place of the all-too-common short-range system approach to staffing systems, in which the fact that there are vacancies to be filled not only constitutes the focal point of the recruitment–assignment process but is also virtually the sole consideration involved, leaving the future of the organization and the interests of the individual largely disregarded.

It goes without saying, presumably, that the data entering into the initially attributed capability–requirements ratio should be kept up-to-date by approved personnel administration and supervisory techniques. Thus the attributed ratio may be appropriately modified from time to time, and the capabilities involved may be targeted toward the most suitable employment.

The principles involved in this recommendation may not differ substantially from some of those taken into consideration in some programs of organization development, manpower planning, management development, and the like, but there is probably a significant difference in both timing and emphasis. The timing recommended is for application at the beginning of the organization–employee relationship. The emphasis is at least equally for the best interests of employee and organization.

Careful consideration of the individual's capability–requirements ratio may not produce the greatest or the most immediate flow of recruits for assignment to vacancies in system staffing. It might even require some modification of assigned tasks, supervision, training, and rotation. But it may well provide the surest and, in the end, the most satisfactory approach to the manning of systems, and pay off additionally through the enrichment of the organization, both in available human potentials and in the constructive nature of individual–organizational relationships.

Capability–performance ratio. It is a generally accepted absolute that no one ever performs consistently at the level of his maximum capability. Although it is sometimes possible under special conditions to induce or to enforce physical exertion up to the limit of capability, performance of tasks other than those purely physical is scarcely subject to the same kinds of constraints. Indeed, there are no truly objective means of measuring nonphysical capabilities. At best they must be inferred from the available indications, which may be inadequate, misleading, or highly susceptible to misinterpretation.

Although the ultimate reaches of an individual's capability remain hidden and untried, it is a common experience of supervisors of system operations that individuals too often perform at a level obviously far below their known capabilities—based upon past performance or some similar objective indication.

We are not referring to the more or less normal day-to-day variation, whereby an employee has good days and bad days, with quantity and/or quality of production varying—often independently—around fairly consistent averages. (This is an important subject by itself, but it is relevant to comment here that narrow deviations and a consistent average indicate the employee is or is trying to become adjusted to the assignment, whereas substantial variations and shifting averages suggest maladjustment.) We are discussing the quantitative and/or qualitative deficiency in performance of a task known or believed to be well within the capacity of the individual.

Many system-related tasks are so standardized and subject to such fixed requirements that the qualifications of individuals to perform them, especially after training, are rather readily determined. The criteria generally applicable to the individual's performance, then, rest relatively seldom on capability and rather generally on the relation to quantitative and qualitative standards.

An individual may not produce up to quota or may not complete a day's work by normal quitting time. An individual may produce more defective work or make more errors than others doing similar work or than his predecessor. An individual may not pull his weight in a group or team. An individual's deficiencies may be apparently attributable to carelessness, lack of attention, or thoughtlessness, usually generalized as lack of interest or poor motivation. An individual may leave an unduly large or unauthorized portion of his work for the night shift or for a senior clerk or for the supervisor.

In many such cases the employee, when interviewed, will frankly admit that the task is well within his power to perform better. He may complain that the volume of work is too great, the opportunity for error too prevalent, or the schedule, arrangements, or other factors too unfavorable. The variety of excuses or explanations is unlimited. (Where groups are involved these often become more emphatic and may take on the nature of grievances.) Complaints about the difficulties or complexities of a task are more likely with groups than with individuals; individuals generally do not like to imply lack of capacity on their part, but are often glad to do so with the support of a group.

There is nothing more futile than to offer all sorts of incentives to an employee in the hope of motivating satisfactory task performance when the task requirements actually exceed his capabilities. On the other hand, it is contramotivational to be involved in a task well below one's capacity. The best performances within a group of employees engaged in parallel tasks often come from those whose capabilities exceed the requirements by the smallest margins.

Supervisors should be concerned about the capability–performance ratios of the employees they supervise and should seek to be realistically aware of the significant indications. This means more than to form a general impression as to whether the individual is performing satisfactorily, or a judgment as to whether the individual possesses potential for advancement or for transfer to more exacting assignments.

The supervisor should be able to develop a relatively clear perception of the individual's capabilities in relation to both the requirements of the assigned task and to his actual performance. And the appropriate resources of the organization should assist, encourage, and require the supervisor to develop these abilities within himself and to apply them constructively.

When an individual's capabilities are clearly adequate to the performance of an assigned task, deficiencies of performance must necessarily be attributed to other factors than lack of capability. Some of these factors may be within the power of the supervisor to identify, and then to eliminate or alleviate. Otherwise, he may be able to alter the task itself or to transfer the individual to some other kind of work under his supervision. Failing this, he should offer the employee the opportunity of transfer to some more appropriate assignment within the organization, and assist through informal as well as official channels in placing the employee more suitably.

It is recognized that supervisors, especially when shorthanded or pessimistic about the availability or quality of replacements, are understandably reluctant to initiate the loss of an employee who is carrying some of the load, and that this reluctance may be emphasized when the supervisor recognizes the employee's superior potential even if he does not and will not apply it. That condition makes this situation an organizational problem; and it becomes a matter of organizational policy to recognize the problem and the potential benefits and to devise programs to dispose of the problem and to secure any obtainable benefits.

Many large organizations, of course, even when not expanding operations, have a chronic need for substantial recruitment, due to turnover. In systems utilizing large numbers of employees with little or only easily learned skills, the ratio of turnover is often high and sometimes recruitment cannot keep up with vacancies. In such situations it is a common practice to mechanize and automate.

The result often is not so much to reduce the number of employees required as it is to lower even more many of the requirements for employee capabilities. In many situations recent high school graduates with no industrial or clerical experience are sought for the many simple tasks. These job opportunities, usually carrying the minimum compensation feasible, are especially sought by those for whom it is relatively difficult to find employment—marginal students and dropouts, members of racial minorities, those with linguistic problems or with alien cultural backgrounds. Observers may note large numbers of such employees performing the less exacting functions in factories and the simplest and most routine of tasks in clerical operations.

The real capabilities of such employees generally go quite unnoted. Their capacity to handle more exacting assignments is commonly ignored. The potentials they represent and possess go to waste.

To the extent that such employees recognize this condition—or have it called to their attention by their leaders or would-be leaders—they will tend to attribute it to factors other than carelessness, ineptitude, or insouciance on the part of supervisors. They will deem it a deliberate policy. And there will tend to be a correlation between the extent of untapped potential of the individual and the intensity and dynamism of his frustrations and resentments.

Measuring Individual Contributions

Quantity and quality of output. The establishment and administration of quantitative norms for production can follow principles and techniques based on generations of experience and deriving from contributions dating from the beginnings of scientific management. Where productivity can be measured in simple units of production, other things being equal, individual contributions can be determined by counting.

Quotas may be established by management, foreman, or union or by the nature of the job. How these are met may be largely a matter for the individual to determine for himself. Obviously, when a uniform quota is applied to all and serves as a limitation on production, the capability of the individual to produce need not exceed the minimum requirement. Where there is little or no opportunity for upward mobility of personnel, this limitation renders the capability–performance ratio largely academic so far as management is concerned.

If a bricklayer may not lay more than twelve hundred bricks per day, and still must be paid for the full day, it may matter little how long it takes him to fill his quota. He may take it easy and finish in eight hours or work faster and go home early. Drillers working in a mine tunnel may be expected to pull a round per paid shift. Once they have drilled the necessary pattern of holes, these are loaded with explosive, and the explosive is detonated, the men are precluded from further work at that face until the gases are cleared. It usually matters little to management if they require a full shift or finish in much less time.

Where piecework or bonus systems are in effect, management or custom provides special incentives for quantitative performance. Payment by the pound for picking cotton or by the ton for mining coal emphasizes the importance of the capability–performance ratio; and if there are large discrepancies apparent between the capabilities and the performance of individuals, the system of incentives should be reevaluated as inappropriate or inadequately effective.

Where the potential productivity of the individual is substantially increased by the provision of equipment, the realization of production potentials gains an added economic dimension. The investment must pay off, and this assumes and requires a volume of production that thus becomes or sets the minimal standard. The productivity of the operator, then, is evaluated largely in terms of the potential production of the machine. So long as this is measured in simple numerical terms, no major difficulty is presented in most situations in setting a standard upon which agreement can be reached in terms of simple numbers.

However, the measurement of production in simple numerical terms assumes equality (or at least substantial or effective equality) of value for each unit. Barring special factors, one pound of cotton is presumably equal in value to any other pound of cotton from the same field; one ton of coal is likely to be of equal value to any other ton of coal from the same vein. But one box of carefully picked strawberries may be worth much more than other, bruised boxes from the same crop; and one ton of steel may be worth much more or less than other tons of steel from other heats of the same furnace. It is this qualitative factor that so greatly complicates so many problems of production and of adjustment of compensation to the value of the economic contribution.

The situation becomes perhaps most clearly uncontrolled and ill adjusted in most clerical operations, especially in those where a number of clerks are performing more or less parallel tasks. No two will be making contributions of truly equal value, not on any one day or even on average. Although quantitatively determined productivity may be equated, the variables introduced by qualitative considerations make the evaluation of the results of individual performance so complex that subjective judgments almost invariably take over. This complexity is emphasized by the fact that in actuality no two errors have the same negative value, and classes of errors are due to different kinds of deficiency of performance.

There are, of course, situations where errors are particularly costly. These may range from waste in the fabrication of expensive materials to impairment of relations with valued customers. Employees who can satisfactorily meet the requirements of such situations are likely to regard freedom from error as something like a primary virtue, and to be distressed at any departures of their own from a virtually error-free standard.

In other situations, by contrast, the emphasis is on the handling of maximum volume at lowest cost, and the inevitability of error must be accepted. Sometimes provision is made for the correction of error after detection outside the organization, as by customers of banks, brokerage firms, and department stores or by the purchasers of lower-grade consumer goods.

There will, of course, be a considerable difference in the sense of responsibility for results, as well as in the nature of the work contributed, between employees working under such disparate conditions as these.

There is generally an adverse relationship between

quantity and quality: The speed required for meeting high quantitative standards tends to increase the ratio of error; the care required to minimize error cuts down on quantitative achievement. Some employees are better suited, by capability and temperament, to emphasize speed in their work and others to emphasize carefulness. When employees are miscast in assignments unsuited to their aptitudes, management may be more to blame for unsatisfactory results than its employees.

Often incentive plans are introduced that tend to produce results other than those desired. For instance, incentives for quantity, unless appropriately related to qualitative standards, may lead to a substantial increase in unsatisfactory production. Unless incentive schemes are properly designed, they may emphasize adverse factors and open the way to disputes.

Too many incentive schemes actually work out in ways that were not intended. The employee may develop a sense of vested interest in doing the job in the way that pays him best. If this deviates from the way that produces best results for management, the scheme is at fault; but it may be difficult to negotiate acceptance of change.

Absences, tardiness, vacations, and other conditions that may reduce the effective size of the work force often have the related effect of proportionately increasing the individual work load. In some cases, this effect can be serious, even overwhelming. And when an absentee's regular daily tasks are not taken care of during his absence, he will return to face an oppressive load of accumulated work—the prospect of which may well spoil a vacation or impair a convalescence.

In many work situations, the work load may vary substantially, sometimes from day to day. This is particularly true of clerical operations that must be completed daily, based upon the load of transactions for the day, despite the fluctuations due to holiday accumulation, first-of-the-month, year-end, and other peak-load situations.

An extra busy day is likely to lead to a greater incidence of errors, but it also imposes, in itself, a greater burden of effort. Some employees can bear this with ease, however protestingly. But others may have real problems and difficulties in coping with more than an average work load. The same principles appear to apply to variations in the daily work loads of service operations and even of some manufacturing firms—especially in job shops and in urgent maintenance situations.

The effect of such painfully supported excess will vary greatly; and so will the reaction to the necessity for working overtime; some employees may welcome the additional pay, while others will resent the invasion of their private schedules.

In systems operations especially, there is a tendency to provide for a high degree of regularity and to regard great fluctuations in the work load as uncharacteristic of well-planned systems. But such assumptions cannot abolish reality, and some systems operations generate inequities in the work loads of individuals.

Management needs to be more aware of and sensitive to such differences, and more equitable in administering adjustments of assignments and work loads. Failure to meet such administrative requirements can only result in intensification of the employees' self-images as victims of the system.

It is logical, perhaps necessary, to evaluate each employee's contribution in terms of the needs of the system and/or the organization. But it is also logical, and perhaps necessary, to evaluate the factors affecting that contribution. The individual's performance is not a clear, precise indication of his capabilities or of his basic disposition. It should also be regarded as reflecting some of the effects of management—intended or otherwise.

Maladjustment of Individuals to Tasks

Work that is inherently unsatisfactory or unpleasant to the worker will of course be avoided if preferable alternatives are available. Such work, therefore, tends to be done by those who have no choice or by those who, at least, prefer it to the alternatives open to them.

Those who have little or no choice are usually those who have the least qualifications and who must compete with others in adequate supply to meet the local demand. These conditions tend to apply to any employees of low seniority, to many urban members of minority groups, and to those with educational, language, and other handicaps. These tend to fill the ranks of the migrant farm workers, the workers at menial tasks, the suppliers of casual, unskilled labor. Another category is the locality-bound worker, as in coal-mining communities, where there are usually few alternatives to entering the mines.

These disadvantaged employees are often conscious of exploitation and resentful of it; and they readily come to the conclusion that the least desirable work should not necessarily be the least rewarded. Leaders and would-be leaders with a wide variety of orientations, motivations, and competence are not lacking. And occasionally, even in what appear to be the most hopeless situations, remarkable progress is made.

One example is the ultimate success of the long-drawn-out strike of California grape pickers (largely Mexican in origin) led by the remarkable Cesar Chavez. Another is the impressive gains of sanitation workers—publicly or privately employed—in many cities, which was dramatized in Memphis by Martin Luther King before his assassination.

An ever broader category would include those in industry who have no special skills. The traditional craft unions did not concern themselves with other workers, except to exclude them from their usually rigidly controlled membership.

The organizers of the industrial unions saw the need for representation of other employees, and thus originated the unions of auto workers, steel workers, mine workers, and others that took in all employees in the plants of an industry. Because this included the lowest levels of qualification, those more highly qualified sometimes held back. There has been a basic difference in interests between the craft and the industrial unions that has not even been successfully bridged by the joining of the American Federation of Labor with the Congress of Industrial Organizations to form the AFL–CIO.

The New Middle Class

As wages went up, standards of living rose, and many workers of little specialized skill are classified today, in terms of earnings, as lower middle class or even middle class. But more leisure, greater job security, and higher standards of living have contributed to increase the resentment of unsatisfactory, undesirable conditions. The employee who lives well is less tolerant of unpleasant environments; he resents noise, smells, and lack of amenities. The better-educated employee, with noneconomically motivated interests, is less tolerant of boredom, repetitive, meaningless (to him) tasks, and mindless activity generally. And he is extremely likely to feel that he should be compensated for what he has to put up with as well as for what he produces.

One result is the widespread and growing demand among industrial employees for an adequate pension after thirty years, providing for retirement at the age of forty-eight for most high school graduates. Another result is the probably increasing incidence of absenteeism—especially on Mondays, Fridays, and after paydays—whereby a growing proportion of hourly paid workers sacrifice wages for the privilege of getting away from their work. This presents a disciplinary problem and can interfere seriously with production.

A third result is seen in the growth of sabotage—deliberate damage to or destruction of equipment and production—by employees deeply resentful of the conditions imposed upon them. Detroit auto plants, for instance, report deliberately scratched body paint, slashed upholstery, built-in rattles, and the like, clear indications of workers' potent, adverse emotional attitudes toward their employment.

The inevitable result of such developments must be to increase the size and the urgency of wage expectations, to exacerbate the bitterness of grievances, and to promote an ever wider range of demands relating to environment and to working conditions. The latter will be reinforced by the growing public concern with ecological considerations.

Although this discussion deals primarily with industrial employment, the same principles apply to employment in clerical systems. However, the intensity of these factors is generally far less apparent because of the lack of organizational means to bring them forcefully into the open and to confront management with the necessity for recognizing the situation.

While the physical environment of the clerical system may have few of the adverse characteristics of the factory floor, the sheer mindlessness of the tasks may be even more stultifying. Suffocatingly repetitive clerical tasks may still require sufficient attention so that the effort to force concentration becomes all but intolerable to some. The conviction that the work leads nowhere and confers no benefits other than pay can occasion both nagging and poignant revulsions.

The inevitable effects of such frustrations include a generalized resentment against everything connected with the job and a hair-trigger conditioning to focus accumulated resentment acutely at any appropriate target of opportunity. Under such circumstances trivialities support irreversibly the conviction that one is irresponsibly or deliberately oppressed, and irrational acts and outbursts become inevitable. First to suffer, of course, both quantitatively and qualitatively, is production. And some resentments, becoming aggressive, seek even more expressive outlets.

Consider a typical employee, perhaps forty years old, working on an automobile assembly line or in a foundry or in any other industrial operation where the physical environment offers little in the way of charm. Twenty years ago he would have lived with his father and mother, brothers and sisters, crowded together in a small home or a city tenement. He would have commuted to work by crowded streetcar. After contributing to the household budget, he would have had little pocket money and his recreations would have had to be simple and cheap.

Now he lives in a two- or three-bedroom house or apartment, drives to work, owns a color TV, and has traveled extensively within (and, perhaps, outside) the U.S. His home is comfortable, and he has learned to enjoy material things. He expects more of life than he dreamed years ago could be his lot. Everything has changed, except, apparently, the place where he works. There he finds the same, or what seems the same, pressure and noise and stressful atmosphere; the same exacting, infinitely repetitive routine requiring the same endless, tiring motions. Because he compares his work time with the greatly improved conditions he now enjoys elsewhere, he finds his job environment ever more distasteful and less tolerable. Because he is now earning more than ever before and probably as much as he could earn in any other way, and because he needs or wants all he can earn, he stays on the job. But if he feels himself helpless to improve conditions, he will still resent having to accept the situation. Securing higher pay helps him to rationalize his position, but does not meet many of his clamorous needs on the job.

Consider also a reasonably alert teen-ager, first employed in a simple clerical position. Once mastered, the job is tediously repetitive and without inherent interest. There are a few older clerks around who seem to have become adjusted to such tasks. He looks at them, forms his own judgment of them, doesn't identify, and is immediately and profoundly convinced that he doesn't want to become like them!

What can he do about it? With no skills or experience, he cannot readily find a better job or higher pay. He stays, but his resistance to the situation can take many forms that indulge his antipathy. Given a union, he may become a militant management baiter. Without a union, he may still be a ringleader of malcontents, a focus of discontent, an initiator of employee demands, a thorn in the side of his supervisor.

Systems too often generate conditions that alienate employees. The indirect cost of such alienation in employee shortcomings attributable to it, and the direct cost in compensatory wage demands can be great indeed, but unfortunately such costs are seldom considered in system design.

The New Generation

It may be worthwhile to examine the strike against General Motors in 1970. The responsible, sophisticated leadership of the United Auto Workers was, presumably, well aware of the consequences of a lengthy work stoppage—to their members, to the company, and to the national economy, which was already impaired by inflation, increasing unemployment, and other ills.

The company made an offer in terms of wage increases and benefits that many competent persons considered reasonable or even generous. The monetary loss to be sustained by an average employee in the course of a strike of the anticipated duration—up to two months or more—would be far greater than the cumulative hoped-for gain in wages could bring in less than several years. And yet the strike was voted. Why?

Some excerpts from the daily press may throw light on the matter. Long before the strike was voted, Leonard Woodcock, then a UAW vice-president, addressed representatives of 225,000 skilled workers out of the 1,400,000 total membership. He declared that the company "lost its old ability through fear and intimidation to impose discipline" and that the company "did not scare people the way they used to."[1]

At the same meeting, Douglas Fraser, director of the union's Skilled Trades Department, acknowledged that absenteeism was a problem the union was also concerned about because it meant harder work for the men on the job. But both Fraser and Woodcock insisted there were "other ways than discipline" to get "cooperation." The younger workers, Fraser said, "have different values than people of my generation. They take very seriously this whole question of individual freedom." He went on to declare that compulsory overtime, in the name of efficiency, infringes on this freedom.

"And we just have to tell these companies, in the spirit of our youth, that the King is dead and we are going to bury him in 1970."

One magazine, evaluating the situation, commented: "Among auto workers, the chief expression of discontent with their lot is a powerful movement for '30 and out'. The slogan sums up the idea of voluntary retirement at any age after 30 years service with a minimum $500-a-month pension . . . 'the number one issue' in this year's bargaining."[2]

The news story continued: "The rank and file aggressiveness arises from new awareness . . . that working conditions have not improved as much as they feel they should have. The assembly line is a noisy, dirty place to work, as it was 30 years ago, and it moves at a fast pace. . . ." The magazine quoted a worker: "No one

[1] *New York Times,* March 10, 1970.
[2] *Time,* September 7, 1970.

wants to go inside the plant in the morning, and no one is sad to leave in the afternoon. You're as much a machine as a punch press or a drill motor. Your life is geared to the assembly line. I've lost my freedom."

The conditions would probably not have bothered his father nearly so much. But now there is a serious clash, for, just as personal standards, values, and frames of reference are becoming more exacting and expectations are higher, conditions in many workplaces have deteriorated. With the progress of automation, tasks have been increasingly subdivided, and many workers are required to perform ever simpler, more repetitive, and less satisfying jobs.

Time quoted an assembly worker: "Do you know what I do? I fix seven bolts. Seven bolts! Day in and day out, the same seven bolts. What do I think about?"[3] He then named a popular sex-symbol motion picture actress. But clerical workers are generally debarred by the nature of their work from such anodynes. They must concentrate on the work sufficiently to discriminate symbolized significances, and to perform accordingly.

The conflict is especially emphasized among the younger workers, whose lack of seniority draws them to the less desirable situations, and who share the general questioning attitude and intolerance of the status quo of their impatient generation. One leader of the West Coast ironworkers, after a 30 percent wage increase, commented: "The younger leadership is not satisfied. I don't understand what they really want, what it would take to satisfy them."[3]

A General Motors plant chairman said that many of

[3] *Time,* November 9, 1970.

the young workers, who had never been involved in a strike, were "raring to go. We're getting a different breed of worker. He's not looking at the dollar so much. He wants more time off and hates working on a line. That 30 and out is a must."[4]

Time magazine gave this evaluation:

> Money alone will not do it. The young workers are revolting against the job itself, or at least the way it is organized. They reject the principle enunciated in 1922 by Henry Ford I: "The average worker wants a job in which he does not have to put much physical effort. Above all, he wants a job in which he does not have to think." The job that has no meaning must often be performed in factories that seem bereft of human feeling. Auto plants are often old, dirty, and so noisy that conversation is impossible. "That's why so many young people just go from shop to shop," says Eugene Brooks, director of labor education at Detroit's Wayne State University. "They can't believe it's this bad. A young guy will start working at Dodge, and after a week he'll be so shocked at how dull the job is and how unpleasant the working conditions are that he'll figure it has to be better somewhere else. So he goes to G.M. for three days, and then to Ford—and then he sees it's all the same. The young guy asks: 'Is this all there is to America?' They're not buying the myth anymore."[5]

Such bitter disillusionments and frustrations occur against a background of general economic, social, and political insecurity compounded of the many pregnant and unresolved issues facing the nation. But, increasingly, the

[4] *New York Times,* September 13, 1970.
[5] *Time,* November 9, 1970.

focus of discontent is the nature, content, and circumstances of the job itself.

The resolution of this problem is essential if the factories and offices of America are to function effectively in the future. And the solutions cannot be developed by social scientists alone. The practical, applicable answers must be produced by competent managers and technologists who are fully alerted to the human realities, and who are determined to meet the challenge by satisfying not only the production goals but also the human needs.

8 The Future of Systems

THE future of systems has overwhelming implications for the future of organizations, and, through them, for the kind of future that awaits all individuals. For the growth, development, and the obvious future potentials of systems promise in no uncertain terms that more and more individuals will be vitally affected directly, by participation, and that ultimately no one will be insulated at least from major indirect effects.

The twentieth century has seen the evolution to a pivotal stage in the Age of Systems that may be termed the second-generation system. This development represents not merely an improved system, but a system developed by systems. This phenomenon, probably signaled most conspicuously by the Manhattan Project, marks the coming-of-age of systems as creators of other systems that are so complex and intricately involved that they could never have been brought into being by individuals working outside of systems.

Such practices as the Critical Path Method (CPM) and the Program Evaluation and Review Technique

(PERT) are typical expressions of the new system-originates-system technology. Supported, reinforced, and given virtually an added dimension by the harnessing of computers to their processes, the advent of Systems-Originate-Systems Technology (SOST) opens vast new perspectives of future systems development that not only try the imagination but also truly defy it. It appears inevitable that systematic projection beyond the known must range far (and more systematically!) beyond the vision of man alone—at least in the realm of the foreseeably achievable.

The information explosion, the knowledge explosion, the age of technology, and the vast, all but unrecognized seismic upsurge of the global sociological tidal wave are all uniting to engulf us in the unimaginable embrace of the Age of Systems, in which, we may be sure, the individual may dispose of now fantastic leverages for specific purposes, but will, at the same time, be more than ever at the mercy of his ever increasingly man-created environment.

Realization of Systems Potentials

The Egyptians had systems for making brick and for raising pyramids. The feudal ages had systems of agriculture. The laissez-faire utilitarianism of the Industrial Revolution had its systems for manufacturing.

All of these systems were more productive (more cost effective) than the practices that preceded them.

All of them met most, if not all, of the standards by which success is measured. They accomplished their primary, direct objectives: They produced the results for which they were called into being. But each of them involved and depended upon the grossest inequalities between those who directed and conducted the system and those upon whose exhausting labor it depended.

Such earlier systems rested upon slavery or serfdom, or upon economic exploitation based upon theoretical freedom of contract that in actuality offered no acceptable alternative. The burdens of such systems were borne by the many; the benefits were enjoyed by the few; and these latter contributed, at most, their crude capital (land, animals) and their autocratic supervision.

The world has not by any means outgrown the organized heritage of such socioeconomic systems. Much of the world's agriculture still rests on conditions that are not far—if at all—removed from peonage and serfdom. For the East Pakistani farm laborer, the Louisiana sharecropper, or the California wetback migrant, freedom of choice and bargaining are almost totally theoretical factors in their situations. Many of the workers in the Calcutta textile factories or the South African diamond fields are little better off. Such victimized individuals suffer the inevitable consequences of receiving a poor share of their own low productivity. Those who manage them generally contribute little but the imposition of their will to compel the performance of simple and well-established patterns of activity.

With the advent of technology and the availability of power sources other than muscle, the potential productivity of the individual rose as did capital requirements and the need for specialized management. This development was paralleled, in the Western-oriented cultures, by a change in governmental institutions tending toward greater participation by the masses and by moves toward the organization of workers to resist exploitation and to create a more effective basis for wage bargaining.

In the more successful and more rapidly advancing economies, the growing demand for workers made collective bargaining more effective. As industrial and agricultural productivity rose in these more advanced lands, the newer trends in government and in labor relations brought about changes in the distribution of wealth that resulted increasingly in higher standards of living for more and more people. Thus the newer systems not only accomplished primary objectives more effectively but they also contributed—incidentally or purposefully—to the benefit of far more of those participating.

Among the inevitable results were a surging of aspirations in a cycle that demanded, first, more education and, then, based on that education, a "better life" than parents and ancestors knew, and, inevitably, more education, more leisure, more aspiration.

Rising aspirations apply to the nature of one's occupation as fully as to anything else. At first such aspirations tended to be satisfied by the achievement of symbols: the office instead of the shop, the white collar instead of overalls, the desk instead of the workbench, the pen instead of the hand tool. But, more recently, there has been an increasing recognition of the differences in the many kinds of work done in offices as well as in factories and laboratories and of the directions in which different kinds of work are likely to lead.

"Jobs with a future" are distinguished from "dead-end" or "nothing" jobs, and this distinction is increasingly important to an ever larger proportion of those seeking employment. As day-to-day living becomes less demanding and more rewarding, the individual finds it natural, and even attractive, to look ahead. Career consciousness commonly begins with the vocational counselor in high school.

The idea of getting ahead in the world is, of course, as old as civilization, but that idea has become a practical one for a larger number of individuals than ever before, and that number is growing. In some countries, and especially in certain identifiable social sectors of those countries, the importance of that idea often transcends that of immediate reward or current status. It governs life perspective and dominates personal planning.

A "better life" has different meanings for different individuals, and may, of course, be measured in many ways and on many different levels. Often concern with the material becomes obsessive, and the quality of life appears to be judged by the visible aspects of the standard of living—impressiveness of home, lavishness of indulgence in preferred activities, and the like.

Satisfaction and enjoyment are subjective, and the actual benefits derived from material possessions and overt indulgences vary with the individual. It seems obvious, however, that many persons seek satisfactions and enjoyments by following examples set by others, rather than by initiating their own deeply desired courses of action. (Riesman has described such borrowed motivations in *The Lonely Crowd.*[1]) The results must often be disappointing, and the individual wonders why he does not enjoy his new pursuits or possessions more fully.

There have always been individuals who pursued a particular activity regardless of any reward visible to others and despite the availability of alternatives seemingly more attractive and rewarding. Such individuals range from scientists to revolutionaries, from professors to poets, from explorers to government officials. They

[1] David Riesman, *The Lonely Crowd* (New Haven, Conn.: Yale University Press, 1950).

have found an occupation, a career, a way of life, work that absorbs all their interests and energies, and they have little interest in anything else. Some doctors, lawyers, architects, and other professionals evince a similar absorption in their work.

Despite the fact that many such persons must tolerate a lower standard of living (by material standards) than many of their friends, neighbors, or colleagues, they are often envied. Others recognize that their dedication brings them intangible rewards of an entirely different order, but apparently of a far more satisfying and less transitory nature than the material gains so generally sought.

It is less commonly observed, but this condition often exists even on relatively humble levels, and even where the performance requirements are decidedly undistinguished. A bookkeeper may find such satisfaction in the accuracy and neatness of his accounts that, to a very real extent, he envies no one. A mechanic may be fully and satisfyingly absorbed in his engines. A farmer may seek nothing more rewarding than the tilling of his fields and the care of his cattle. A carpenter may love his work. Undoubtedly, many individuals at any economic level find sufficient satisfaction in the performance of their daily tasks so that economic considerations are secondary or, at least, not dominating.

Looking at individuals in another way and from another perspective, we see that some have a decided tendency toward acceptance of and adjustment to different situations, while others find this difficult and tend to resist, object, and complain. Some people are easily satisfied; others are never satisfied. Such variant dispositions may, indeed, be due to the effect of primary or major needs. If such needs are met, others may be of no great importance. And if these needs are not met, the satisfaction of other needs cannot be substituted.

In any one case, it is impossible to determine to what extent the individual's satisfaction or dissatisfaction with his way of life is due to a general disposition toward satisfaction or dissatisfaction, or to a real coincidence of or conflict between existing conditions and personal values, and, if the latter, to what extent the relation between the individual's value system and his occupation plays a part.

Contributions of Systems to Participants

Whatever the reactive "mix" in the case of any one individual, it would appear certain that industrial and clerical systems will be increasingly judged and evaluated by the attitudes of participating personnel. The most potentially effective of systems will be increasingly at the mercy of those who staff it, for the fullest realization of its capabilities.

Those who seek interesting, satisfaction-producing jobs with a future will shun other jobs, and there will be an increasing proportion of such highly selective job seekers. Any such who find themselves, for lack of alternative, compelled to work at less satisfying jobs will increasingly suffer frustration and will increasingly react resentfully.

In the industrially advanced lands of Western orientation, it seems clear that any occupation that is inherently disagreeable or unsatisfying is somehow out of tune. Socioeconomic history clearly shows that, in the U.S., the dirty work has been passed on, from generation to generation, not so much from father to son as from one racially and linguistically distinctive wave of immigrants to another. There has not been one hereditary caste of "untouchables," or a "permanent lower class," or a static role for all "hewers of wood and the drawers of water" (though no doubt blacks have had more than their fair share of relegation to the least desirable occupations).

The expansion of population, the westward migration, the unprecedented development of agriculture and industry, the availability of free land and of rich, unclaimed wealth in natural resources all coincided to bring about both unlimited upward mobility—economic and social—and a growing need for labor.

In the earliest settlements of the Northeast, men did their own work, or grouped together, or exchanged labor. Soon there was an infusion of indentured servants, Negro slaves (freed, in the North, before the Civil War), and the penniless. The potato famines brought the Irish, who largely dug the canals and built the railroads—supplemented by Chinese at the western end. As the Irish moved upward, the Central Europeans and the Italians took over their picks and shovels, moved into the multiplying steel mills, foundries, mines, slaughterhouses, and any other work available to unskilled men who could not speak the language. (The Jews were not a major factor in the labor force, except in a few trades, and they were especially attuned to upward mobility.)

During and after World War I, the immigrants and their offspring began to move upward and to be replaced, in the less exacting tasks, by Negroes moving up from the South—unskilled, uneducated, and handicapped by generations of exploitation and neglect. Since that time, there has been no major wave of immigration to move in at the bottom except, perhaps, the Puerto Ricans, who have tended to stay in a few big cities, and the Mexicans, who are to be found mainly in the Southwest. (The Cuban refugees from Castro have demonstrated an impressive upward mobility.)

One result of this near stratification, coupled with a relatively high level of employment, has been the spectacular increase in levels of compensation for many forms of work that are inherently unpleasant and unsatisfying. The handling of refuse, the dismantling ("wrecking") of old buildings, and similar unattractive tasks command wages on a level with or superior to those obtainable from many more congenial pursuits. Formerly, much of the most unpleasant work of the world was done at the lowest wages, by those having no alternative.

But now the relatively high rate of compensation for the performance of tasks requiring little skill or experience is a clear indication of the increasing necessity for offering special inducements for the undertaking of work that many would find unacceptable. Since many would find such tasks unacceptable at almost any price that could be reasonably envisaged, the work is done by those who do find the compensation adequate for the unpleasantness. Such inducements are less and less effective as personal standards develop and the opportunities to indulge them become available. This is illustrated by the increasing scarcity of full-time household servants at almost any level of compensation.

Staffing Problems

These observations mean that managements will find it increasingly difficult and costly to staff systems that require individuals to perform work that is not inherently satisfying—work that is tedious, uninteresting, dull, unrewarding. And already we see the clearest evidence of this accelerating trend in the rapid turnover in most system operations among the brighter, better educated, more aspiring recruits.

In many situations, most of the longer-service employees, who have been able to make more or less satisfactory adjustments to their tasks, are identifiable as deriving from family and socioeconomic backgrounds that leave them with a perspective so conditioned that their present calling enjoys a satisfactory rating. They are inured to tedium, have largely satisfied any aspirations for learning and advancement, and enjoy emoluments and status that are significantly superior to those enjoyed by their parents or achieved by most of their cultural peers.

It is no mere coincidence that an ever higher percentage of the longer-term employees in large clerical systems, for instance, are members of minority groups, and derive from ethnically, linguistically, and educationally disadvantaged backgrounds. But one is compelled to ask, how large is the reservoir of employables having acceptable levels of competence who retain sufficient tolerance of the conditions to staff such stultifying positions? And how large will this reservoir be in the future, especially in relation to the ever growing requirements?

Technology has demonstrated an extraordinary resourcefulness in providing substitutes and leverages for human activity. Until recently the human activities thus replaced or facilitated were largely of a physical nature; power machines have taken over or improved upon the widest variety of previously manual functions. Many of the manual functions involved in clerical activities have also been mechanized or converted to electronic processes, and thus myriad repetitive tasks have been eliminated.

Now computers are increasingly taking over functions that were once mental. But competent clerical workers are still generally in short supply, and the younger and less experienced among them seem seldom to regard a job as permanent. And the capacity of the business world to generate requirements for more paperwork seems constantly to outrace the capacity to handle it.

Personnel Require Changed Practices

Technology has presented the world with fully automated manufacturing systems of many kinds that produce standard units—whether kilowatts or engine blocks, filaments or pills, metal parts or plastic items. Control of quality—however difficult or complex—is facilitated by the condition of standardization in which only deviation from a common standard need be noted.

But clerical systems are fundamentally different. An essential characteristic of clerical systems involves distinguishing the symbolic differences that make each item significant. Made up as they are of a limited number of common symbols (primarily letters and digits) each item is unique due to the selection, combinations, and varying quantities of these symbols appearing therein.

"Accuracy" can be common to all clerical items only as an abstract quality. Accuracy requires each item to be different from all others, but in a specific way, directly related to the distinctive content of data or information to be conveyed. Furthermore, while manufacturing standards are expressed within stated limits of tolerance (that is, plus or minus 1.25 millimeters), no such latitude is permissible in most paperwork, where exact amounts (especially of dollars) are a sine qua non.

While certain useful—even essential—processing of paperwork can be entrusted to machines, it seems impossible to eliminate some degree of human intervention at both input and output stages, at least. Whether increasing automation of clerical processes will actually reduce personnel requirements is indeed problematic. So far, the expanded volume and the profusion of applications have consistently outpaced the capacity to process.

We have discussed the foundations for certain basic assumptions:

1. There will continue to be a growing requirement for the processing of paperwork—whether it is called clerical work, bookkeeping, accounting, data processing, management information, or whatever. Existing needs and applications will continue to grow in volume, and many new applications will be developed.

2. The increasing capabilities and capacities of machines will not alone meet these requirements. (One may also assume that the many and resourceful suppliers will continue to evolve improved and more versatile equipment and services. While this must inevitably contribute to filling part of the gap between needs and resources, past experience suggests that it is likely to increase the proportion of unexacting, near mechanical tasks to be performed by humans. This assumption should continue to be valid, despite substantial technological breakthroughs.)

3. There will be a continuing and probably increasing growth, establishment, and utilization of large systems for the handling of paperwork.

4. These systems will have a continuing and probably increasing requirement for personnel.
5. Relatively fewer employees will accept long-term involvement in the lower-level tasks similar to those called for in most present clerical systems.

If these assumptions prove to be even partially justified by developments, the solutions so generally sought to present-day problems of mass clerical processing will prove increasingly impracticable and ineffective in the future.

If those facing an increasing requirement for large-scale paperwork processing cannot find their solution in more productive equipment and/or in an adequate supply of undemanding and tolerant labor, they must seek salvation in the resources that are available to them. Optimum utilization of the most effective equipment should afford some advantage. But this is most likely to exacerbate the growing difficulties with personnel—if handled according to presently prevalent practices.

The obvious, logical corollary to this proposition is to change present practices realistically, taking into account human values and psychic costs as well as organizational and task requirements. This will require not only drastic changes in recruitment, selection, indoctrination, orientation, and job training but also in supervision and in system management. Changes will probably also be advantageous in the character of assigned tasks, in the working environment, in the form (and probably the magnitude) of compensation, in the organizational structure, in career planning, and in other areas.

If the constructive potentials of all such reforms are brought to bear, then the ingenuity of the designers of systems and of systems equipment may be permitted to bear fruit. But if those who plan systems fail to plan adequately for the indispensable human involvement, the future of systems will be increasingly clouded by deficiencies, adversities, and other problems due to the failure to provide for compatibility between the specific needs of and for the functioning of systems, and the physical and psychological needs of the people who are indispensable to their operations.

Evolving systems may indeed bring to the family of man benefits that will make subsequent stages of the Age of Systems far better times than any that have gone before. But before that can come about, the system must make its peace with the people who enable it to function.

Realization of Organizational Potentials

When an individual faces a simple task made up of similar units of work, the scheduling of specific operations and the overall planning required may be greatly simplified, even obvious. But the dimensions of the task—the total number of operations to be performed, regardless of similarity or repetitiveness—are translated into time. Thus it may take a reasonably skilled axman two hours to split a cord of kindling, one log at a time. If he is joined by another, of equal skill, they may reasonably expect to divide the duration of the task in half. If one man can dig a ditch in eight hours, four men of equivalent capability should be able to do it in two. And so on.

The arithmetic for such tasks is simple and straightforward, and the organization for such parallel tasks is generally also quite simple. The basic function of supervision, in such situations, calls primarily for seeing to it that each individual contributes something like an average share of the total task.

To meet this requirement only, the ratio of supervisors to the supervised can be low. A foreman of a railroad construction gang in the pick-and-shovel days often bossed over a hundred men. He concerned himself with little besides insuring that no one escaped his proper quota of work. He was little or not at all concerned with the personalities, problems, aspirations, or potentials of the men he bossed, and he had little or no qualifications, inclinations, or time to consider such matters.

The earliest forms of men's group activities—hunting and fighting—brought about the organization of the tribe so that the generally similar and parallel activities of the men were coordinated to achieve a common goal. Each individual contributed directly and shared in the results, and so had a reward as well as a sense of achievement when the results were successful.

When all the people of a seaside village join to pull in a great fishing net, as in some Philippine villages, each has a sense of being a part of the life-supporting operation and each shares in the catch—thus having a direct, personal interest in the results.

But in most systems the reward for participation is quite indirect, and results are often obscure to the worker at contributory tasks, which are frequently extremely simple, repetitive, and common to a group.

Where simple, identical behavior patterns are required for each participant, the deviant is generally conspicuous, and one pair of eyes can adequately oversee the significant actions of many, be they in a Roman galley or a Macedonian phalanx, a cotton field in Mississippi or a line of Caribbean stevedores carrying bananas on their shoulders. In close order infantry drill, the sergeant major can tell if any man in the company is out of step or place.

Participants in such group activities may at best derive satisfaction from, perhaps, only three sources:

1. *Membership in the group.* Individual attitude toward group membership is a measure of morale, which may be high or low, corresponding to a positive or negative attitude toward membership in the group.
2. *Performance of the task itself.* Satisfaction from the performance of a task itself (apart from achievement) is derived from the social and physical environment and the physical or mental experience of the operation. Repetitiveness, monotony, and tedium are adverse factors for most (but not all) individuals.
3. *Achievement.* This may be qualitative or quantitative. It may also be absolute or competitive, that is, derived from meeting or exceeding a standard or from relative accomplishment among fellow workers. The meeting of an easily attained standard of performance offers little satisfaction, once achieved. And standards generally attained by the majority are often raised. The satisfaction of superiority must be limited to the few.

Thus it seems clear that, where common tasks are performed by all the members of a group, the psychic rewards—the satisfactions from acceptable performance—are extremely limited.

An apparently inevitable concomitant results from this basic organizational truism: The extent of supervision provided tends to be limited to that minimum required to insure the meeting of at least minimum standards of performance by the members of the group. This condition alone places the supervisor under the necessity of treating all workers having the same tasks in much the same way, modified primarily by the need for attention to those violating or failing to meet standards of performance of task.

Those employees meeting such standards, according to this implicit plan, require little or no attention. Thus supervisory consideration tends to be allocated in inverse proportion to the value of the contribution of the individual employees. This condition, almost literally, obtains in many systems, both clerical and industrial. And too high a ratio of supervisees to supervisors (in relation to the nature of the task) makes it all but inevitable. "The squeaking wheel gets the grease."

The consequences of this, of course, are such that each employee will sense his individuality to be slighted, since the supervisor concerns himself solely with task performance. Furthermore, such conditions make obvious a close correlation between supervisory attention and unsatisfactory performance, with all that this may imply for day-to-day sensitivities. As the harassed and overburdened supervisor ignores the better-performing employees and spends the available time with the poorer-performing individuals, his attempts to upgrade them often involve adverse criticism, admonition, and—often—expressions of exhausted patience, and threats (implied or specific) of disciplinary action.

The supervisor is thus a taskmaster rather than a leader; and regardless of his qualifications and inclinations, he may have little choice about this under the conditions imposed by the organizational format within which the system is required to function. But the psychological consequences of such situations must be adverse, cumulatively; and they will affect not only the employees but also the supervisor and the organization itself.

Among those who have recognized the growing unacceptability of tedious, repetitive tasks, there has been talk of and some considerable experimentation with the introduction of some degree of variety and change into the task assignment of the individual. In a typical "job-enlargement" project, for instance, workers in the assembly of an electric typewriter were taught to perform a considerable number of steps instead of a single step.

Many such experiments have been abandoned, because they tended to operate precisely in opposition to the principles of successful mass production. And, indeed, when manipulating tangible items, such as parts of an assembly, the productivity of a worker must necessarily be measured in much the same way as the productivity of a machine.

The advantages of job enlargement, if any, must be measured in terms of intangibles or possibly in terms of overhead factors, if they can contribute to a reduction in costs related to employee turnover.

The quality of production, in manufacturing, is usually subject to immediate inspection or test, and the application of appropriate and (usually) economically feasible quality control procedures.

But job enlargement in clerical operations often involves very different factors and considerations.

The introduction of variety or complexity into a clerical task may bring about consequences not easy to foresee or to control. And yet the opportunity often exists, and (unlike most manufacturing) may even carry with it the possibilities of efficiencies and cost reductions.

As an example, many clerical systems involve daily routines that depend upon other activities outside the system, and that require different activities as the day progresses. For instance, the inputs may be found only in the incoming mail or in certain deliveries or transmissions, and the processing follows, in stages. This condition is characterized by batch processing, rather than constant flow processing. It may thus be feasible for some employees to work on each of several different or successive stages of the processing of a daily batch.

In the case of systems requiring relatively small staffs, the whole system (or much of it) may be handled within a single working group, with individuals shifting from one phase of the processing to another. Properly handled, such situations can offer some individuals a degree of satisfaction not obtainable from concentration upon a unitary task.

When an individual faces a complex work activity, he should know how best to schedule his time and efforts and how best to apply his resources. The determination of such matters is generally termed planning. When it is done for an individual, it necessarily casts him in the several roles required by each component task, and thus assigns to him the various functions and operations required to fulfill each task.

The necessary training must be provided, and presumably the extent of this will be considerably greater than would be required for a single task. In many cases, this increased requirement can best be met by teaching one task at a time, rather than by attempting to teach several tasks at once—which may be confusing. With the learning of one task, the trainee can become productive, and this alone should add a more favorable context to the learning of additional tasks. Such a gradual introduction to the prospective functioning of the individual also offers a desirable opportunity to observe and evaluate the individual, with a view to fostering adaptation to the tasks and to the group or, if indicated, making a change in assignment.

Systems and the Organization

In the Age of Systems the organization must be adapted to the effective exploitation of the systems upon which it depends for its most efficient functioning. Actually, most organizational designs inherit the traditional hierarchical structure of a less systematic epoch, and often incorporate systems within the vertical compartments representing line or staff departments. An exception is increasingly found in the case of information systems, which are often set up more or less independently, but usually report, at or near the top, to a key officer (frequently the controller).

The principal drawback to such an arrangement is the simple but unavoidable fact that the executives at the higher levels above systems operations are generally placed in such positions (or the administration of the system is placed in such positions under them) for reasons other than the desirable one—that these executives know best how to operate the system successfully over the long term. The placement of systems operations under a senior executive is too often dictated by organizational considerations other than those related to the problems of administering the system.

The principles guiding the usual organizational arrangement include the executive's need for information developed by the system, or his overall responsibility for the function that the system is intended to serve, or perhaps merely that, in the narrowing concatenation of top-level population, he is an obvious choice, or the best available for the purpose.

The System Manager

Under such circumstances, there is little recourse at the lower level of the manager of the system for competent assistance in administering the human aspects of his charge. Since he himself is primarily (and necessarily) a technician, he usually needs help in optimizing operations where people are concerned. But his superiors are in no position to provide this help.

He may, if sufficiently conscious of the problems arising from the disparity between operational and human needs, turn to the director of personnel. However helpful this important staff officer may be, he will be reluctant to attempt to introduce substantial exceptions to the policies already accepted and approved by top management, and especially so if he foresees that these will involve substantially increased burdens for his already fully extended staff. Yet the only hope lies in more intelligent, more emphatic, more truly competent handling of the administration of system personnel.

Blueprint for Personnel Well-Being

Imagine that the recruitment and placement processes are so perfected that each candidate for assignment to a system is truly well adapted to the post under consideration. Imagine that the candidate, fully informed of the realities and the possibilities, makes the choice freely, among respectable and relevant alternatives. Let the neophyte then be thoroughly indoctrinated with a complete understanding of the way in which the tasks he is to perform fit into and contribute to the other phases of the system, and the manner in which the system serves the overall organization.

Let the working group be so molded by the leadership of its supervisor that the newcomer easily relaxes into a pleasant, cooperative relationship with his fellow workers; and let the supervisor also train, guide, counsel, and encourage the recruit so that he gains confidence and quickly achieves a respectable competence.

As he masters the assigned task, let him be made aware that his progress is noted, that his potentials have been evaluated, and that his progress to another task or post is being weighed. Let him realize the significance of this as the nature, requirements, and circumstances of possible transfers are made familiar to him. Let the appropriate transfer take place not too long after the time is fully ripe. (This progression is intended to produce an experience of a very different order from that imparted by nearly automatic "merit" increases, and other impersonal practices.) And let him again experience the same helpfulness and encouragement as he begins to master the new responsibilities.

Such attention to and fostering of the careers of employees takes much of the time and effort of competent, interested supervisors and supporting staff, and must rest upon the understanding by senior management that organizations are made for people, and not vice versa. Perhaps, in some situations, specialists will be required who are best qualified to administer to the human needs of particular systems.

But the results of such practices, which can be extremely valuable now, will, in all probability, be ultimately essential, if organizations are to operate successfully, in the Age of Systems, with employees who seem, somehow, not too well adapted, or ready to adapt, to those forms of employment that are increasingly characteristic of this phase of the age.

The Future of the Individual

It has been widely recognized in recent years that the individual who is working at tasks below his potential is likely to be dissatisfied, and so may perform unsatisfactorily and cause other problems. Long before, observers noted that some people performed tasks unsatisfactorily because they were apparently incapable of better performance. Faced with both of these undesirable alternatives—too little potential and too much—managers have usually chosen to err on the side of choosing excess capability. This decision has been justified by recognition of the need to provide for the future of the organization by promotion from within and by the sound policy of promoting, on merit, those who have demonstrated the potential for more exacting and more responsible assignments.

The principles involved in these considerations have not changed, but, as a result of the proliferation of systems, their importance may have been ignored in too many situations.

Systems impose rigid requirements. The tasks involved in system processes are relatively fixed; they are often extremely simple and repetitive. Jobs outside of systems can often be varied to fit the individual incumbent, and standards of performance also can often be adapted to the individual, but with systems, such adaptability is seldom feasible. This inflexibility does not allow for the needs or aspirations of the individual. Rather, it imposes an uncompromising environment, and the effect on attitude is generally for the worse, since the individual must conform, in some degree, at least, to the rigidities of the situation.

If adjustment and adaptation were not so unilateral, there could be greater scope for fitting functions and tasks to individuals. And if individuals received well-conceived guidance, assistance, and support in making their adjustments to their assigned positions, no doubt their acceptance of conditions—and resultant attitudes—would improve. Furthermore, if assignments were better adapted to individual capabilities, interests, aspirations, self-images, and the like, then the difficulties of adjustment would surely be diminished.

No one seriously questions that technology, and perhaps especially cybernetics and automation, can and increasingly will increase worker productivity so that all will gain and have much more, by working far less. The parallel development of systems could presumably enhance both productivity and leisure.

But who knows or can even make convincing predictions about the effects of all such "progress" on mankind? Or on its more privileged individuals? Or on those, perhaps inevitably, far less privileged?

If the importance of the work in one's life is related in some degree to the time it occupies, then what will be the effect of a workweek of thirty hours or twenty or ten? Will this mean a greater willingness to support tedious routines or unpleasant environments or nasty tasks, since there are fewer hours in which these must be tolerated? Or will the heightened expectations and enlarged horizons, greater familiarity with amenities, and increased habituation to comfort, pleasure, and esthetics —will such inevitably make the less satisfying occupations increasingly unacceptable?

As individuals acquire some degree of independence from the overwhelming necessity of providing for their own survival, they begin to turn their interests, and sometimes their energies, to matters not directly related to the satisfaction of survival needs. When there is a choice of what to do and what to consume, preferences develop. As the range of choices widens, tastes form, grow, and in cultural and recreational pursuits as well as in sensual indulgences, tend to become fixed. With habituation (which does not necessarily take a long time), the indulgence of a preference often becomes a matter of substantial importance, sometimes motivating drastic behavior.

An expense-account salesman in Chicago may stir up far more of a fuss over an unsatisfactory martini than a hungry dweller in Calcutta will make over missing a badly needed meal. A diligent college student will deeply resent a disturbance to his studies that may be welcomed by his roommate. A suburban husband sacrifices some of his sparse opportunity to be with his family in order to lock himself into his den and watch football on TV. Workers walk off the job because a foreman has offended one of them. A housewife with none but household duties opens cans or heats frozen foods rather than allow more elaborate cooking for her family to interfere with bridge or television.

Who can make confident predictions about the future tastes, recreations, hobbies, avocations, interests, and aversions of those whose time will be increasingly freed from the pressure of employment? Will they apply themselves to extend their education? Will they listen to music, collect art, study literature, follow ballet? Will they paint, compose, write, dance, philosophize? Will they turn to politics, civic endeavors, philanthropy, social services, charitable works? Will they concentrate on self-development or on serving their neighbors? Will they never tire of golf or fishing or shuffleboard?

If we observe some of the examples set by those who were or who made themselves economically free of regular employment, we behold the most extraordinary extremes. We see some of the independent sons and daughters of the wealthy dedicating themselves wholeheartedly to the service of the underprivileged, while others withdraw completely from the establishment into communes or into other forms of hippie or antiestablishment environments, where they engage in behavior that appears to revolve around such pivotal attractions as uninhibited (and often multiply addressed) sex indulgence and the use (or abuse) of narcotic drugs of various degrees of clinical destructiveness and irreversible addiction.

We see the wives of well-to-do men, freed by servants and schools of most chores of household and child rearing, who turn to charities, services to the community, cultural activities, or other pursuits generally regarded as beneficial to others or self-improving. And we see others in similar circumstances who devote themselves to the round of pleasures—be it bridge, gossip, sipping cocktails, shopping, extramarital involvements, or just plain idling, with or without movies, matinees, television, or light literature "pour passer le temps."

We behold tens of thousands of retired persons, including many who have held positions of considerable responsibility and who have compiled admirable records of accomplishment. We see them trying to adapt to an endless vacation, and we see them too often bored with the unceasing recreational round and discouraged by the unlimited prospect of more of the same.

When we think of the vast benefits to the individual that come with greatly increased leisure, should we not question how such leisure is to be spent, utilized, savored, lived through? Will not this be an important aspect of the future of systems?

Clearly, the privileges of leisure and of economic independence may be put to an extraordinarily wide range of uses.

But there are other problems, also, in this connection, some of them deeply rooted in our economic system, our system of distribution, and other fundamental structures of our economy and culture. Among these are some of the phenomena that have provided the enemies of capitalism with their least answerable objections. If the organizations and systems of the past, and perhaps of the present, do indeed support and give occasion to such objections, then perhaps some care should be taken with the systems and organizations of the future that such abuses of the past may be minimized.

A common characteristic of the bourgeois and mercantile evolution has been that of the individual so skilled in the manipulations of the marketplace that he accumulates a fortune far beyond his conceivable needs, and continues to accumulate when there can be no understandable related satisfaction other than the mere fact and experience of accumulation. The contributions of such individuals to the economy from which they draw so much are usually not nearly so apparent as the rewards or, at least, the material withdrawals that they have been able to garner as a result of their activities. To many, the value of what they have taken unto themselves appears highly disproportionate to the value of what they have contributed.

How will this disproportion, if such it is, be resolved as the Age of Systems brings ever greater leverage to the effective self-seeking of individuals skilled in the enlargement of their share of the wealth produced by the application to growing sectors of the economy of ever more potent systems?

Another common, even prevalent, characteristic of at least the earlier (and current) decades of the Age of Systems is manifested by the individual with special competence in the political and social skills that so often facilitate advancement through corporate hierarchies, and that seem often to be reinforced most effectively by a single-minded determination to rise within the organization. The successes of such individuals too often appear to be achieved despite, or even partly because of, lack of scruple or of serious regard for any considerations other than their own advancement. Their progress may seem to many due far more to connivance, cronyism, personal leverages, and other such factors than to such considerations as merit, competence, and potential for contribution to the effectiveness of the organization. What effect must the ever increasing impact of systems have upon this common organizational phenomenon?

It is, of course, perfectly feasible to adopt a relatively circumscribed perspective. Accustomed as we are to thinking in terms of improving our systems and perfecting our organizations, we are quite capable of going on in the same way, assuming that our goals of higher efficiency and greater productivity stand justified on their face value and that the prospective results of easier work and more leisure must be regarded as major benefits.

But we have the clearest evidence that easier work and more leisure may not always prove so beneficial. In fact, we may observe a regrettably high proportion of instances in which these are or may have been deleterious, or at least have been unmistakably associated with and appear to have causal effects upon or to have contributed to developments and forms of behavior in individuals for which we would prefer not to accept the responsibility.

It would seem, then, that those who would shape the future of systems, and thus of the organizations that increasingly embody and utilize them, must also concern themselves with the future of individuals.

Participation in the functioning of systems can be an effective learning and conditioning experience, as participation in a working group can teach all sorts of lessons that are formative in effect and may prove highly applicable elsewhere (for better or for worse).

The effects upon individuals of participation in systems must increasingly become not only a matter of concern for those who shape the conditions that will influence the workers but also the subject of constructive planning and provision that will help, increasingly, to fit them for a future to which they can realistically look forward without alienation. Expertise in and responsibility for systems must increasingly incorporate long-term consideration for the development of individuals if the Age of Systems is to bring lasting benefits to man.

Index